TREES ANCIENT AND MODERN

CW01497284

TREES ANCIENT AND MODERN

Woodland Cultures and Conservation

CHARLES WATKINS

REAKTION BOOKS

Published by
Reaktion Books Ltd
2–4 Sebastian Street
London EC1V OHE, UK
www.reaktionbooks.co.uk

First published 2025
Copyright © Charles Watkins 2025

All rights reserved

EU GPSR Authorised Representative
Logos Europe, 9 rue Nicolas Poussin, 17000, La Rochelle, France
email: contact@logoseurope.eu

No part of this publication may be reproduced, stored in a retrieval system or
transmitted, in any form or by any means, electronic, mechanical, photocopying,
recording or otherwise, without the prior permission of the publishers. No part
of this publication may be used or reproduced in any manner for the purpose of
training artificial intelligence technologies or systems.

Printed and bound in Great Britain by Bell & Bain, Glasgow

A catalogue record for this book is available from the British Library

ISBN 978 1 83639 118 0

CONTENTS

Choose therefore, trees which nature's hand has sown
In proper soils, and climates of their own.
RICHARD PAYNE KNIGHT, 'The Landscape', 1794

Forests were set on fire – but hour by hour
They fell and faded – and the crackling trunks
Extinguish'd with a crash – and all was black.
LORD BYRON, 'Darkness', 1816

There is not an acre of forest that has not its significance;
not a clearing, not a thicket that does not suggest analogies
to the labyrinth of human thoughts.
HONORÉ DE BALZAC, *Le Curé de village*, 1841

INTRODUCTION

S ome people find forests threatening and troubling: indeed, woodland clearance has often been celebrated as a sign of increasing
prosperity and a surrogate for civilization. But most today are concerned about the loss of woodland and might agree with Lord Byron's
Childe Harold in wishing

> To slowly trace the forest's shady scene,
> Where things that own not man's dominion dwell,
> And mortal foot hath ne'er or rarely been;

trees here providing a solace preferable to the solitude to be found
in 'the crowd, the hurry, the shock' of society. It is generally agreed
that existing woodland should be protected from clearance and conversion to other land uses, that culturally and ecologically important
older woodlands should be protected, that active management should
be encouraged, and that new woods should be established on surplus
agricultural land.[1]

Trees and woods are deceptive. The largest trees in a wood are
often not the oldest. Small seedlings and saplings under the shade of
larger trees may remain almost moribund for many years until conditions for them improve, perhaps following the loss of the shading
trees from disease or storm damage. In gardens and public spaces trees

that appear to have been growing there for many years may have arrived very recently. Examples include the thousands of ancient olive trees, readily available from nurseries, which now adorn private gardens, corporate headquarters and public squares across Europe.

Trees hold memories of distant and more recent events. Ronald Blythe, when contemplating in late autumn his oak trees dating from around 1800 and planted for timber for Nelson's navy, looked forward to the time when all their leaves would fall:

> And then, wonderfully beautiful and youthful though they are in full dress, stripped naked they reveal the pitiful lacerations of the great 1987 gale. Broken boughs, hang limbs, wounds like those under the bright uniforms of naval officers, are bared for all to see. Every year I wait and hope for the winter winds to bring them down, but still they swing high up against the sky.[2]

The links between trees and memory are usually hazier and less literary and high flown than this. The diarist Francis Kilvert left his parish of Clyro, Radnorshire, on 2 February 1872 and recorded

> The day I came to Clyro I remember fixing my eyes on a particular bough of an apple tree in the orchard opposite the school and the vicarage and saying to myself that on the day I left Clyro I would look at the same branch. I did look for it this morning but I could not recognize it.[3]

Although deceptive, trees have long been used as markers of time. In 1426 at Kidderminster, Worcestershire, in a legal proof of age enquiry, Robert Whateley 'testified that he remembered the baptismal day of the disputed heir, because an oak tree growing in his orchard had been "totally blown to the ground by a strong wind"'.[4] The death of the tree was as memorable as the birth of a child.

Woods, forests and trees are rarely out of the news, good and bad. They have become enormously popular and are often identified as unproblematically beautiful and benign. The enthusiasm for and love of trees brings with it deep concerns that they are fragile and threatened. Carbon sequestration, which has only recently been widely recognized as a vital benefit of forestry, has now become a leading factor influencing public opinion and government policies about trees. The clearance of tropical forests to produce timber, palm oil and soya beans causes deep worry. Moreover, devastating and deadly fires in California, the Amazon, Siberia and Australia emphasize the fragility of forests. Fires are a major threat in Europe too; they are less extensive, but equally potent in Portugal, Greece, Croatia, France and Italy.

Europe is a heavily wooded continent with 46 per cent forest cover, but there is much variation from country to country. Foresters have argued strongly for more trees since the seventeenth century, when woodland cover reached its nadir. The twentieth century saw the doubling of woodland area in England from 5 to 10 per cent, and in the UK to 13 per cent. This was achieved through state forestry and government support for private woodland owners through grants and tax concessions, but it remains very low compared to the area of woodland in the EU (40 per cent) and to countries such as France (32 per cent), Germany (33 per cent) and Italy (33 per cent). The countries with the largest proportion of woods and forests are Finland (74 per cent) and Sweden (69 per cent).[5]

There is a broad consensus today concerning the need for an increase in European woodland. In the UK, for example, people worried about climate change, nature conservation, public access and health issues have joined foresters to demand more woodland. All political parties agree that there is a need to establish large new areas of woodland to help absorb CO_2. The aim is to increase UK woodland cover from 13 per cent to 19 per cent by 2050: still very significantly less than the EU average of 40 per cent. There is much less consensus about where the woodlands should be established, who should own and manage them and what form they should take.

The interactions between climate change and forestry are complex and uncertain. Trees are not just passive respondents to climate change; they actively reduce levels of CO_2 in the atmosphere. Much research remains to be done to discover which tree species and forms of woodland management are best at absorbing CO_2. It appears that quick-growing conifers absorb more than deciduous woodland. A recent report by Forestry Commission researchers indicates that planting a hectare of lightly managed broadleaf woodland would result in an average CO_2 uptake of 1.3 tonnes CO_2 per year until 2050, and that a hectare of moderately growing coniferous woodland would soak up 4.9 t CO_2 per year for the same period. The important contribution that surviving fragments of ancient woodland across Europe can make to the carbon-carrying capacity of the continent has also been emphasized.[6]

Although woodlands cover almost half the surface area of Europe, their interactions with humans and animals remain mysterious. Unlike much agriculture, where the annual cycle of cultivation, planting and cropping is easily comprehended, the life cycle of most trees is longer than ours, and they gain over time diverse cultural attributes which vary from person to person. Forests have competing cultural claims and understandings. Woods have varied dramatically through time and from place to place. This book explores these issues with examples illustrating distinct forest histories and tree stories. European trees and forests cannot be considered apart from American, Asian and Australian trees, which are linked through history by introduced species and trade. Europeans have been enticed by novel introduced trees since ancient times. Pliny was fascinated by the origin of trees and was especially keen to learn which countries they came from: 'foreign trees' were an essential element of his tree classification.[7]

Fascination with trees and woodlands is increased by the immense diversity of the landscapes which they create. The word 'forest' conjures up images of places dominated by trees and wild-ness, of sylvan nature untouched by the hands of humans. But the reality of forest history is much more complicated. There is no direct connection between the idea of forest and the concept of woodland: medieval forests were legal entities more like modern national parks than dense, extensive areas of managed trees. In many parts of Europe forests are sites where the competing interests and demands of vil-lagers, aristocrats, farmers and monarchs have been played out over centuries. Rather than natural woodlands, they are places dominated by the management of wild and domesticated animals, and by the commercial control of trees. But they are also places of myth whose landscapes and legends have provided inspiration to poets, novelists and painters.[8]

The relationship between humans and trees is complex and the many different values ascribed to trees, woods and forests provide them with rich layers of association and meaning. Trees and woods often outlive humans and can provide a semblance of order, continu-ity and security. Woodland management across Europe has changed constantly over centuries in response to changing economic and social demands. Knowledge of many ancient practices, such as coppicing, pollarding and shredding, is rapidly being lost. There is a considerable body of research on the cultural and social history of different tree-scapes, and their value for landscape and nature conservation. Much of this is interdisciplinary and brings together historical, social, cul-tural and scientific knowledge. The work of historical ecologists has been particularly effective in influencing the practical management of treescapes such as wood pastures and ancient woodland. There has also been much recent interest in the relationship between trees, art and conservation. The long life of trees means that they often create 'lagged landscapes' resulting from previous cultural, social, economic and political activities. The original reasons for woodland establish-ment may have disappeared, but the trees, now with a patina of diverse and wide-ranging values, associations and meanings, continue.[9]

During the last glacial maximum (LGM) in Europe, around 25,000 to 18,000 years ago, there were extensive ice sheets over large parts of northern Europe reaching down to the English Midlands, eastern Denmark and northern Poland. Those parts of the continent just to the south of the ice consisted of mixtures of grasses, herbs and low shrubs, possibly with some patches of birch (*Betula* spp.) woodland, where mammoths and woolly rhinoceroses grazed and helped create an open steppe landscape. Further south in Europe, pollen analysis indicates that there were extensive areas of woodland. In the Iberian peninsula, for example, there was a mix of pine and broadleaved woodland and open parkland. There was considerable continuity in woodland cover during the LGM in central Europe. A study of the vegetation history of the Carpathian (Pannonian) Basin indicates that 'most of the present-day native flora (1404 species, about 80%) can occur in climates as cold as or colder' than the LGM. Indeed, the researchers argue that 'long-term continuity of much of the flora in the Carpathian Basin is more plausible than regional extinctions during the LGM followed by massive postglacial recolonizations.'[10]

As the climate warmed, and the human population grew, trees and shrubs spread northwards across the continent. It is likely that around 8,000 years ago approximately 60–70 per cent of Europe was wooded. Since then woodland has been fragmented and reduced to around 30–40 per cent through conversion to other land uses, primarily arable land and pastures. Pollen scientists have shown that there were major differences in the histories of mixed temperate woodland across much of Europe and the northern coniferous forests. The mixed deciduous temperate woodland started declining some 6,000 years ago and this was strongly associated with woodland clearance for agriculture. In the coniferous northern forests, by contrast, the impact of human activities only became 'detectable in the last two millennia and has left a larger area of forest in place.'[11] At a more local scale, there was considerable variation in the timing and extent of the loss of long-established woodland. In Poland, for example, pollen evidence suggests that there was some forest expansion from the fifth to the ninth century AD, after which there was 'large-scale forest decline, especially in north-western and north-central Poland'.[12]

There are no parts of European woodland that remain untouched by human influences. This is true even of the mountain birch woods that grow in northern Scandinavia between the coniferous forests and the treeless northern alpine heaths. These are thought by many to be 'virtually devoid' of human influence, with their upper tree lines 'driven by climate and natural disturbance' with little historical woodland clearance. But a study of the birch woods of the Adamvalta Valley in Sweden, just inside the Arctic Circle, indicates that the felling between AD 800 and 1200 of 'mountain birch trees for fuel and wooden constructions may have deforested an entire mountain valley within just a few hundred years' and that prehistoric clearance could be 'both dramatic and persistent' even after a short period of tree felling.[13]

In this book I explore the cultural geographies of British and European woods and forests, and to an extent in the USA, focusing on key themes that exemplify human interactions with trees. I draw on a wide range of sources and approaches, including literary descriptions; drawings, paintings and photographs; surveys of present-day plants and animals; historical maps and documents; dendrochronology and oral histories. Advances in understanding the genetic history of tree species emphasize the complicated interplay of humans and other animals in the spread of trees around Europe. The book examines the different ways in which people have used, understood and appreciated woodlands and the implications for current and future forest management and policies.

Chapter One starts by introducing some of the cultural implications of the naming of trees and woodland. It then goes on to consider the invention of the category of ancient woodland as a device to encourage its conservation and management. The chapter also explores the fascination many people have with ancient trees. Chapter Two examines the craving for novelty in terms of the modernization of forestry practices and the associated eradication of old trees. It then considers the cultural context of the introduction and use of foreign trees to make novel and extensive European commercial forests in the nineteenth and twentieth centuries. The main examples discussed are eucalyptus in the Mediterranean, pines in Croatia and Sitka spruce (*Picea sitchensis*) in Scotland and Ireland. The chapter concludes by assessing the current enthusiasm for rewilding former agricultural land, often making use of the natural regeneration of mainly native tree species.

Chapter Three considers the complexity of the relationship between animals, trees and woods by looking at four animals that have been introduced or reintroduced to the UK and many parts of Europe: wild boar, woodland pigs, beavers and Reeves' muntjac deer. Chapter Four examines the threats posed by tree diseases and fire to European forests. It focuses on the introduction and impacts of three of the many increasingly common and virulent diseases: Dutch elm disease, ash dieback and chestnut blight. The chapter concludes with an assessment of uncontrolled forest fires and the potential benefits of controlled burning and grazing as ways of reducing fire risk. Chapter Five explores diverse attitudes to tree felling, including the horrified response of Gerard Manley Hopkins to the cutting down of a row of poplars in Oxfordshire and the enthusiastic tree felling exploits of William Ewart Gladstone. It also examines the attitudes of Charlotte Mew and Thomas Hardy to tree felling and woodland management and concludes with a consideration of the role of women in woodland management.

Chapter Six investigates cultural perceptions of the links between trees and human well-being. It opens with a consideration of Walt Whitman's therapeutic use of trees and then examines E. M. Forster's deep affection for trees, including Piney Copse, his own small Surrey woodland. It explores how palm trees came to provide a cultural signature to the Italian Riviera in the nineteenth century and the recent intense controversies around the felling of street trees in Sheffield.

Chapter Seven examines the sacred nature of trees such as the Glastonbury thorn and the cedar of Lebanon, both in Lebanon and England since the seventeenth century. It goes on to consider the relationship between William Wordsworth and memorial pines in England and Rome. The Conclusion reflects on the implications of the different understandings of trees on future European trees and woodland.

1 Martin Hoffmann, *Carl Linnaeus in Lappish Dress*,
18th century, oil on canvas.

1

ANCIENT ORIGINS

The naming of trees has long been controversial and continues to be so. Different languages have distinct histories of naming trees, enmeshed in local and national cultures. The botanical Latin names allotted under the Linnaean system are an Enlightenment attempt to aid identification, and avoid confusion, when trees and other plants were increasingly traded around the world, especially from the sixteenth century onwards. Carolus Linnaeus (1707–1778) was fascinated by naming plants from childhood. The family name Linnaeus itself was introduced by his father and commemorates a huge lime or linden tree that grew at the family home at Råshult, in southern Sweden. As a student he travelled widely in northern Sweden, Lappland as it was then known, collecting plants from 1732 to 1735. An undated portrait by Martin Hoffmann (illus. 1) shows him 'in Lappish dress' holding a named plant, *Linnaea borealis,* and a vasculum or plant-collecting box. He first published his enormously influential *Species plantarum* in 1753. In this book Linneaus attempted to name all plants consistently, with a twofold name giving genus and species, and allowed botanists to identify, classify and categorize them.[1] For most trees the binomial, botanical name is fairly straightforward and uncontroversial. The first part of the name *Alnus glutinosa*, for example, is derived from the Latin for alder, 'Alnus', and refers to the genus. The second part, 'glutinosa', refers to the slightly sticky nature of young stems and leaves and provides the species name.

As the number of plants known to Europeans grew very rapidly through the eighteenth and nineteenth centuries, botanists increasingly used the names of explorers, geographers, botanists or sponsors that they wished to thank for funding plant-hunting expeditions. And, as with colonial territories named after European towns, cities and monarchs, they were often named to curry favour with powerful political figures. This could become especially important when a tree was identified by those in the nursery trade as one likely to become a valuable commodity. The complex background to naming can be seen with the tree now known as *Sequoiadendron giganteum*. It was first reported to a meeting of the California Academy of Science in San Francisco by Augustus Dowd, a hunter, in 1852. The English plant collector William Lobb attended the meeting and rushed to Yosemite to collect seeds and shoots. The following year young trees were offered for sale by the Exeter and London nurseryman James Veitch for £3 2s each and gardeners and foresters around the world were very keen to plant the tree. Early tourists were fascinated by the tree and soon wanted to see it growing in California. Among these was the indefatigable traveller Marianne North (1830–1890), who painted a view of the Calaveras Grove in 1875 (illus. 2). By then the facilities for tourists were well established and she stayed for a week 'in the comfortable hotel under the big trees'. This was 'indeed a luxury, to be able to stroll under them at sunrise and sunset without any delay or trouble'. She was, however, very concerned about the felling of redwood forests and wrote that 'it broke one's heart to think of man, the civiliser, wasting treasures in a few years' that had previously existed unharmed for centuries.[2]

But what should the tree be called? According to the tribal chairman Floyd Franco Jr, speaking in 1992, the tree was known by Native American members of the Tule River Tribe as 'toos-pung-ish Hea-miwithic'.[3] In Britain the tree was named *Wellingtonia gigantea*

2 Marianne North, *The Calaveras Grove of the Big Tree or Wellingtonia, in the Evening*, 1875, oil on board.

in 1853 by John Lindley in the *Gardener's Chronicle*. It seemed to him highly appropriate to name this huge tree after the Duke of Wellington (1769–1852), defeater of Napoleon and prime minister, who had died in 1852, the year of its collection by William Lobb. In America it was given the name *Washingtonia californica* by C. F. Winslow in 1854 in the *California Farmer and Journal of Useful Sciences*, named of course after George Washington (1732–1799), America's first president. There were several later namings, including *Taxodium washingtonianum* (1854), *Sequoia wellingtonia* (1855) and *Gigantabies wellingtoniana* (1866), before its current name, *Sequoiadendron giganteum*, was finally established in 1939 by the botanist John Buchholz in the *American Journal of Botany*. Irrespective of these changes in the Latin name, the species is still commonly known and marketed as Wellingtonia in the UK.

The naming of the genus *Sequoia* is itself a matter of controversy. It was named by the Austrian botanist Stephan Endlicher (1804–1849) in his *Synopsis coniferarum* (1847, pp. 197–8). It has been thought since the 1870s that the tree was named after the Cherokee Sequoya (George Gist; *c.* 1770–1843), who created the Cherokee syllabary. This is currently contested and there is a lively academic debate.[4] Recent research suggests that sequoia woodlands 'are likely to be the most effective forests worldwide for carbon sequestration', as they accumulate above-ground biomass 'at a faster rate than any other vegetation'. The trees also remain highly productive throughout their long lives and 'are resistant to disturbance such as fire'.[5]

One of the consequences of Latin binomial names being published by non-native botanists is that 'numerous plant names commemorate persons alien to the country where the plant occurs, a practice that continues today.' In addition, 'people commemorated in plant names were often figures associated with colonialism itself.' Several botanists now argue that the cultural and political advantages of changing names may outweigh any potential confusion. This is especially true of names that are 'particularly disagreeable in that they commemorate ruthless dictators or imperialists'. The authoritarian Portuguese prime minister António de Oliveira Salazar (1889–1970), for example, remains commemorated in the Angolan species *Kalanchoe salazarii*, even though his name has been removed from Portuguese bridges and streets.[6]

In addition to removing names celebrating controversial individuals, there is also a move to include more indigenous names in

3 Albert Percy Godber, *The Lone Kauri Tree on the West Coast Road,*
North Auckland, c. 1915–16, photograph.

taxonomy. The New Zealand ecologists Shane Wright and Len Gillman argue that this is important 'for the affirmation of Indigenous Peoples' knowledge of species and their ecology'. They recognize that 'indigenous names enjoy an inevitable priority given that they usually pre-date, not just a non-indigenous taxon name, but the Linnaean system itself.' Such name changes could also 'have positive outcomes for biodiversity conservation due to the potential for increased engagement by Indigenous Peoples'. They note that the Linnaean nomenclatural structure 'provides a simple and elegant way to view evolutionary relationships at a glance' and that the flexibility of the Linnaean system means that indigenous names can be substituted within it by 'swapping a single component'. Taking the New Zealand kauri pine as an example, the current name is *Agathis australis* and the species epithet 'australis' (meaning southern) 'could be replaced by the indigenous Māori name of kauri to become *Agathis kauri*' (illus. 3).[7]

Inventing ancient woodland

Traditional tree management practices such as coppicing, pollarding and shredding almost died out in Europe towards the end of the twentieth century. Coppicing remained an important part of rural economies until the end of the nineteenth century, and lingered on in some areas until the 1960s and '70s. But it then became a largely forgotten practice, and areas of overgrown coppice in some parts of Europe by the 1940s and '50s were likely to be classified as 'scrub' areas, the name scrub indicating that they were uneconomic and hence ripe for clearance for agriculture or conversion to plantations of broadleaved and exotic species. The traditional methods of coppicing were largely forgotten within two generations. In the twenty-first century there has been an increase in the demand for firewood and coppicing is once again recognized by many people as a sustainable and economic form of woodland management.

In some parts of Europe there is great interest in the study of what are now termed 'ancient woods', those thought to be in existence for around five hundred years, which frequently support larger numbers of species of plants and animals than more recent plantations. I am old enough to remember when ancient woodland did not exist, in the sense that it was not a term commonly used by British woodland historians and conservationists until the 1980s. In fact it had been used earlier by foresters but had fallen out of use by the end of the nineteenth century. It then lay dormant until the 1970s, when it was reinvented by historical ecologists to help protect and conserve semi-natural woodlands, which had managed to survive through the first three-quarters of the twentieth century but by the 1960s were threatened by conversion to coniferous plantations or clearance for agriculture.

The idea that many woods originated as modified remnants of natural woodland was accepted by many nineteenth-century foresters and land agents. James Main argued that historical and geological evidence indicated that 'the greater part of the continent of Europe, as well as its islands, were at an early period almost entirely covered with wood.' He thought that some tracts of forest had been preserved in royal forests and private parks and other areas 'of natural forest are also in existence, occupying broken or marshy ground, or precipitous slopes inaccessible to the plough'. He recognized that remnants 'of the aboriginal oak [*Quercus* spp.] woods are still met with in many places, and they are of various character, according to the management they have received'. Various terms were used to describe such woods. They could be 'aboriginal', 'natural' or 'ancient'. One of the first to use the term 'ancient woods' was the Lincolnshire land agent J. West, who in 1842 made a clear distinction between 'ancient woods', which mainly consisted of coppice or coppice with standards, and plantations. The term 'ancient' was also used in the New Forest Act of 1877, which gave protection to the 'ancient and ornamental' woods found in the forest.[8]

For much of the nineteenth century it was relatively rare for old woodlands to be replanted with conifers. The Bagshot nurserymen John Standish and Charles Noble, in their descriptions of newly arrived exotic trees in *Practical Hints on Planting Ornamental Trees with Particular Reference to Coniferae* (1852), enthusiastically point out that 'It is often the object of proprietors to remove woods which are composed of the ordinary indigenous trees of the country, and to replace them with others of an exotic and more ornamental character.' But realistically, they considered that 'the advantages of such existing woods are generally too great to allow their removal.'[9]

Old coppice woodlands retained several advantages in the nineteenth century. There was no need to replant cut areas with expensive nursery stock, the poles produced were variable in size and could be used to make many utensils and tools as well as for fencing poles and firewood, and coppicing allowed income to be produced on a regular basis. For example, hop poles took between fifteen and twenty years to grow and hazel thatching spars only two or three years. There were also important continuing links between specific industries such as tanning and oak coppice.[10] The economic historian Ted Collins considers that there was a Golden Age for the coppice-with-standards system and traditional British woodlands between 1750 and 1850.[11] In some regions of the UK coppice prices remained high later in the century but in others there was a significant decline. Linked to this there was also an increase in the planting of conifers such as larch (*Larix decidua*) and Norway spruce (*Picea abies*) in old woodland to act as 'nurses' to encourage the growth of high-quality broadleaved timber.

By the middle years of the century it became economic for old woods to be grubbed up for agriculture. John Evelyn Denison, an MP and landowner who was Speaker of the House of Commons (1857–72) and president of the Royal Agricultural Society (1856–7), noted in 1855 that 'A few years ago there was a great fall in the value of woods in this district not only as regarded the prices of timber and bark, but of spring wood also, as the cultivation of hops, which had afforded a market for the ash-poles, was generally abandoned.' His friend Thomas Huskinson, land agent for the Epperstone estate, Nottinghamshire, provided the example of Brockwell Hill Wood (illus. 4), of 136 acres on fertile loamy soil, which 'consisted chiefly of oak timber, and ash [*Fraxinus excelsior*] and hazel [*Corylus avellana*] underwood. It had been wood for a long period; certainly for 500 years, and probably much longer.' He had no doubt about the age of the woodland and specifically identifies it as 'ancient', adding that, though 'the stock of timber' was not as great 'as many of the ancient woods' in the county, it remained profitable. There were around 38 oak trees per acre and the 'underwood was of excellent quality: it had been cut at intervals of 18 years, and being in the vicinity of a good market, realised higher prices than the average of woodlands'. In 1840 the owner decided to grub up the trees from 'this wood and convert it into arable land'. Huskinson calculated that the clearance and conversion to agriculture produced an increase in income to the owner of £323 per year and that there were 'few cases in the county where a pecuniary gain would not result from converting woodland into tillage'. Denison

4 George Sanderson, *Map of the Country Twenty Miles round Mansfield*, 1835: detail showing Brockwell Hill Wood, Epperstone.

reported that in 1852 he had also cleared 100 acres of woodland from his Ossington estate and found in addition to the financial gain that it was difficult to overestimate the benefits of 'the removal of the heavy shade of the wood and the constant dampness that hangs about it, from the admission of the sun, and the free circulation of air'.[12]

By the start of the twentieth century, prices for coppice wood, apart from in hop-growing areas such as Kent and Herefordshire, had fallen so low that the influential forester Sir William Schlich, a key figure in the introduction of Continental and especially German forestry methods into Britain, thought that 'taking coppice woods as a whole, their value has fallen so much that in many cases the produce is actually unsaleable.' For him the time had come to consider their future role in the rural economy. If they were used for the preservation of game then 'coppice in combination with standards may still be indicated, but in all other cases, coppice woods should be converted into high forests.'[13] However, most landed estates were hard-hit by the late nineteenth-century agricultural depression and were in no position to start large replanting schemes in their old woodlands, nor to convert such woodland to agricultural land. It was in this period that the term 'ancient wood' disappeared from the literature: they were increasingly seen as worthless and lost their voice and identity. They survived through anonymity and lack of management.

What did early ecologists think of the old 'natural' woods that foresters and land agents were busy forgetting and ignoring in the early twentieth century? How widespread was the understanding that some areas were modified remnants of natural woodland? We can examine this by considering the way that contributors to volumes of the *Victoria County History* written between 1900 and 1911 referred to woodland.[14] Only five authors make direct reference to the likelihood of any woods being 'primeval' or 'aboriginal'. Those who did so used botanical evidence. The botanist George Clement Druce in 1902 emphasized the difference between recent and old woodland in Northamptonshire. He thought that many of the woods consisted of 'blackthorn [*Prunus spinosa*] thickets, or plantations of small trees, and of larch, which make excellent game and fox coverts, but have a singularly unvarying lower vegetation'. It was 'only in the remains of the older woods . . . that any great variety of woodland plants is to be found'. Augustine Ley, writing of Herefordshire in 1908, was much more explicit in stressing the importance of old woodland. He emphasized that 'the rich development of the fruticose Rubi in Herefordshire' was mainly due 'to its heirloom of the aboriginal woodland'. He thought that 'when contrasted with the poverty of the planted "spinneys" of Leicestershire and other English counties' this was strong evidence 'that the Herefordshire woods are really aboriginal'.[15]

Thus botanists, through a detailed consideration of the lists of plant species occurring in different woods, were able to deduce that certain woods were what we now term 'ancient'. But this was no simple task and the geologist Clement Reid, writing in 1899, argued that 'woodland plants are extremely difficult to deal with, partly on account of the wholesale destruction of ancient forests' and partly due to large-scale planting with trees from other areas, which had 'profoundly modified our woodland flora'. It was not only the greater variety of species found in old woodland that interested botanists; they were also fascinated by the curious distribution of certain plants. Reid was intrigued that certain plant species only occurred locally, although there were many other places where the conditions would enable them to flourish. He thought that these 'anomalies are particularly difficult to understand' because of the 'altered state of our woods'. He argued that it was necessary to study 'each patch of ancient woodland, however small', for it was by 'searching small isolated patches of old forest' that it would be possible to identify outliers of fauna and flora which 'once extended over wide areas' now used for farming or housing.[16] Here Reid not only specifically uses the term 'ancient woodland' but suggests a project of woodland research which was to come to fruition eighty years later in the form of the Ancient Woodland Inventory.

Throughout the first 75 years of the twentieth century there was a hiatus in the use of the term 'ancient woodland'. It disappears from the literature. A group of early ecologists including Arthur Tansley formed the British Vegetation Committee in 1904 to survey and map vegetation. This committee became the British Ecological Society in 1913.[17] They wanted to establish whether there remained any 'natural woodland' in Britain or whether 'existing woods' had been 'so altered by planting and in other ways that they no longer represent the native plant communities, but are rather to be considered as mere congeries of introduced and indigenous species'. They concluded that the 'great majority' of British 'natural woods as opposed to plantations' could be classed as the 'lineal descendants, so to speak, of primitive woods'. These 'semi-natural woods, though often more or less planted ... are characterised by many of the species which inhabited them in their original or virgin condition'. They thought that the survival of such large areas of old woods was due to 'the innate conservatism of the English landowner, as well as the backward state of forestry practice in this country'.[18] Tansley later tended to emphasize the artificiality of the remaining semi-natural woods and the difficulty of identifying them, noting in 1939 that 'we can rarely distinguish with any certainty between those which have been planted, and those which are the direct descendants of natural woodlands that have been continuously exploited for many centuries.'[19]

The devastating impact of the two world wars in the first half of the twentieth century and the formation of the Forestry Commission, whose function was to reinvigorate forestry practice and inculcate scientific forestry practices throughout the country, resulted in great pressure to increase the output of wood and timber from older woods. However, it was not until the rise of government subsidies for plantation forestry, especially after 1945, that old woodland began to be planted up with conifers on an extensive scale. In the post-war period the standard forestry texts remained silent about coppice, which was no longer profitable. Even worse, it was seen as a dangerous threat to the growth of young conifers planted in woodland felled and devastated during the Second World War.

In 1949 the practical forester B.R.G. Hammond made a study of how long it took for coppice shoots to die off under a plantation of vigorous beech (*Fagus sylvatica*) trees in southwest England. The beech had been planted in an 'old hardwood stand on chalk downland'. Great efforts were made to try and ensure that the coppice regrowth did not harm the young beech trees. The management policy was to 'weed hard from the start' by cutting back the coppice shoots twice a year. He noted that 'Many foresters will know the great amount of weeding required to establish a forest crop; and during the first three years on old hardwood areas, it is safe to say all of them will be troubled with coppice shoots.' He thought 'the vigour such stools can show' was 'remarkable' and found that 'oak from main-crop stools, ash from young vigorous coppice stools', birch and hazel were 'all troublesome up to the fifth year'. This suggested 'in theory' that 'after five years the crop will generally have suppressed, or the weeding will have killed, the coppice.' But he thought that in practice it was unlikely that all young coppice shoots will have been removed and hence 'we have the insidious tree coppice creeping in.' How could the 'power of coppice to withstand cutting' be fought against?[20]

A study from 1955 provided a partial answer. A forester reported what he had learnt from the 'rehabilitation of devastated woodlands' such as areas of oak standards over mixed coppice growing on boulder clay in the East Midlands that had been felled in the Second World War. He found there were two main types. First, 'Those having a full crop of coppice with heavy grasses and bramble fast disappearing'. Here the best method was to allow the coppice to grow on until it was saleable and high enough so that when thinned it would cast light shade, allowing 'oak planted underneath to come away'. The second type was 'Those having scattered slow growing coppice together with blackthorn, briar and bramble and an extremely heavy growth of grasses'. These areas required a 'smothering crop' such as

Corsican pine (*Pinus nigra*), Douglas fir (*Pseudotsuga menziesii*) or Norway spruce, 'which is easily the favourite', and 'although slow to start it will completely suppress the grasses and coppice.' Overall, he concluded that 'there was no easy way of rehabilitating these woodland areas.' The most effective approach appears to have been to smother the coppice regrowth with the dense shade of conifers.[21]

A revival of interest in the idea of woodland with a link to some form of natural, original woodland was instigated by two foresters, H. M. Steven and A. Carlisle from the University of Aberdeen. In their study of Scots pine (*Pinus sylvestris*), *The Native Pinewoods of Scotland* (1959), they brought together pollen evidence and ecological and historical approaches to woodland history that had normally been kept apart. The evidence they gathered showed 'that the surviving native pinewoods are the lineal descendants' of early post-glacial woods. They thought that 'to walk through the larger of them gives one a better idea of what a primeval forest was like than can be got from any other woodland scene in Britain.'[22] But the real turning point in the modern invention of ancient woodland was the result of research by a group of scientists from different disciplines who developed the approach of historical ecology in the 1960s and '70s.

A key figure was Oliver Rackham (1939–2015), who with George Peterken agreed to use the term 'ancient woodland' to describe the remnants of semi-natural woodland. Rackham's first book, *Hayley Wood: Its History and Ecology* (1975), was one of the earliest detailed assessments of the history and ecology of an individual wood. Rackham defined what he meant by 'historical ecology' in the introduction to his influential book *Trees and Woodlands in the British Landscape* (1976). He pointed out that Joyce Lambert's *The Making of the Broads* (1960), which had demonstrated that the Broads were the result of human activity and not natural features, 'was a pioneer work in historical ecology, taking into account vegetation as a third dimension in a historical and archaeological synthesis'. Since then, he argued, it was only with studies of hedges by those working at Monks Wood

Experimental Station, such as Max Hooper, that historical ecology had been developed 'on anything like a national scale'. Rackham stressed the importance of the historically informed work undertaken by Colin Tubbs in the New Forest, Ruth Tittensor on the Loch Lomond oak woods, and George Peterken. In his own research and study of woods of eastern England he emphasized the importance of rigorous field survey, careful attention to the precise distribution of plants and trees in particular places, and linking this distribution to historical sources, manuscripts, maps and buildings. In his magnum opus *Ancient Woodland* (1980) he summarized his approach: 'Historical ecology seeks to interpret the natural and artificial factors that have influenced the development of an area of vegetation to its present state. It is the history of particular woods . . . rather than the history of generalizations about woodland.'[23]

The reinvention of ancient woodland was of enormous importance in stimulating research into woodland history and conservation. It came in the nick of time because in the 1960s and '70s, as we have seen, old semi-natural woods were often classed as worthless scrub that was uneconomic to manage. Government subsidies favoured their conversion to plantations of quick-growing conifers. Smaller semi-natural woods on farms were threatened with clearance and conversion to agricultural land, also supported by subsidies. A technical report commissioned by the Ministry of Agriculture, Fisheries and Food in 1957 on 'the reclamation of derelict woodland for agricultural use' argued that there was a large area of 'unproductive and uneconomic' woodland that 'if not put to some other uses will remain derelict and a harbour for pests'. Results from a Hampshire study area indicated that reclamation 'can be a most profitable undertaking' and that 'many farmers with isolated areas of derelict woodland on their holdings' should be encouraged to clear it.[24] The impact of these threats was quantified by the Nature Conservancy Council's (now Natural England's) Ancient Woodland Inventory (AWI), established in the 1980s, which, using mainly map evidence, listed ancient and semi-natural woodland for the whole country. It indicated that around 35,000 ancient woods survived but that most of these were small. Since the Second World War, when around half a million hectares of ancient semi-natural woodland survived, about 10 per cent had been cleared and converted to another use. Around 30 per cent had been converted to plantations, and about a half remained largely unmanaged. Only 10 per cent survived 'under the traditional form of coppice, wood-pasture and high forest management'.[25]

By being redefined as ancient woodland these forgotten, tired and largely unmanaged old woodlands gained a new character. They became worth fighting for, and local conservation groups and charities started to purchase them and manage them using traditional methods, including coppicing. The woods started to increase in capital value, becoming a valuable attribute of farms and estates, frequently mentioned in estate agents' sales particulars. The AWI is currently being updated and a new category of 'Long Established Woodland' that has been present since 1893 has been established. The government has a specific policy 'to recognise the value of England's ancient and native woodlands and ancient and veteran trees'.[26]

The importance of ancient woodland for the conservation of biodiversity is increasingly being recognized across Europe. The ecologist Monika Wulf emphasizes the value of mapping ancient forests in Germany, where just over three-quarters of all woodland has been classed as ancient forest that has been wooded for more than two hundred years. Perhaps because so much woodland is classed as ancient, 'Germany still lacks laws or guidelines that specifically address the protection of ancient forests.'[27] A study of semi-natural forests in Hungary since the eighteenth century showed that they had declined to 36 per cent of the total forest area by the 2010s, the reduction caused mainly by conversion to agricultural land. The authors conclude that it is 'immensely important to distinguish between semi-natural and secondary forests, and between types of continuity' when planning for conservation.[28] In Turkey Simay Kırca and colleagues have demonstrated the cultural and conservation values of ancient woods of box (*Buxus sempervirens*) and yew (*Taxus baccata*) and the cultural values of ancient black pines (*Pinus nigra*) on Sandras Mountain.[29]

A recent French study of five mountain and sub-Alpine protected areas emphasized the importance of 'ancient forests', which had been in existence for at least 150 years, for the conservation of woodland species that have 'low dispersal and competitive abilities' and hence have difficulty surviving after woodland is destroyed. The authors argue that although the total area of temperate woodlands in Europe is increasing, the area of ancient forests is still declining. Indeed they estimated that '10 to 40% of the forests present in the nineteenth century have disappeared in European countries today.' They note that only two countries, the UK and Belgium, have specific policies to protect ancient forests and argue that the conservation of ancient forests should be given a higher priority than policies to encourage afforestation, in order to ensure that threatened forest species are conserved effectively.[30]

We can see the impact of the identification of ancient woodland on attitudes to an individual tree species by considering the strange case of the wild service tree (*Sorbus torminalis*), one of the least-known trees in Britain and Europe (illus. 5). It has a widespread distribution from Spain, England and France in the west through Italy and central Europe and the Balkans to Ukraine and the Caucasus in the east. Its leaves may easily be mistaken at first sight for those of a maple. It can grow into a large tree but has traditionally very commonly been coppiced and is often disguised in the woodland understorey. John Evelyn in 1670 identified it as a profitable coppice crop. He argued that when you 'espy' a 'Ches-nut, Service, or like profitable Tree, clear it from the droppings and incumbrances of other Trees, that it may thrive better: Then as you pass along, prune and trim up all the young Wavers.' He recommended that when these wavers or coppice shoots were old enough to be harvested care should be taken to cut them no more than 6 inches above the ground and ensure that the stumps

5 *Sorbus torminalis*, the wild service tree, in flower, May 2014.

were protected from browsing cattle.[31] The wood could be used to make charcoal and as a fuel, and its timber had specialist uses, including making the screws of wine presses. Its fruit, sometimes called chequers, was in the nineteenth century 'brought to market both in England and France; and, when in a state of incipient decay, it eats somewhat like that of the medlar'. Although rather pleasant when over-ripe, they are rarely eaten today but can be consumed as a form of grappa made in Italy.[32]

One of the few tree enthusiasts to recommend planting wild service trees was Uvedale Price, who wrote to his friend Lady Beaumont on 9 February 1805 offering to send her at Coleorton, Leicestershire, 'some trees very common in our woods that may not be so' in hers and which 'the nursery men may not keep for sale, + which you may be glad to have, such for instance as the wild service: I mention it, as it grows remarkably well under shade'.[33] By the late nineteenth century the wild service tree was largely forgotten in England and Wales. In 1906 it was noted that it 'never seems to be gregarious' and 'being usually looked on as underwood and not allowed to grow up to its full size, does not attract notice, and is unknown except to the most observant woodmen, even in districts where it occurs'. It did 'not seem to have any value as a forest tree, and is rarely procurable from nurserymen in this country'.[34] By the mid-twentieth century it was described as 'rather local in oakwoods and ash-oakwoods'; 'handsome' but 'nowhere apparently very abundant in a wild state, and is rare also in cultivation' and in Herefordshire 'rather rare'.[35] Richard Mabey considered the wild service to be 'one of the most local and least known of our native trees', while in Worcestershire it has been described as 'easily overlooked, particularly in shady woodland, since it seldom occurs anywhere as more than small, scattered groups of trees'.[36]

The fortunes of this tree changed strikingly from the 1980s onwards as it became recognized as one of the few tree species that was a remarkably reliable indicator of ancient woodland. This is because in Britain and cooler parts of Europe the tree rarely naturally regenerates from seed and was only occasionally planted by foresters in the last three hundred years or so. The tree is rare in central European forests but 'can persist and regenerate in dense, closed oak coppice forests, presumably mostly through root suckers'.[37] Individual trees have survived usually by being coppiced, and groups of trees develop from the tree's habit of sending out suckers that grow into new trees when not overshadowed by other trees. Consequently, local populations have survived for hundreds of years in isolated ancient woodlands but are rare in secondary woods. With the rise in interest in ancient woodland conservation in the 1980s the wild service tree gained celebrity status and has been classed by the Botanical Society of the British Isles (now the Botanical Society of Britain and Ireland) as an 'axiophyte': a plant 'of particular interest to botanists, conservationists and ecologists', which is helpful for determining conservation priorities.[38]

The increased popularity of the wild service tree means that trees are increasingly being planted to add diversity to secondary woodland. A species that for hundreds of years languished in underwood obscurity has achieved almost cult status among conservationists. The trees are very easily grown from their suckers: I have grown several in this way in hedges and new woodland over the last thirty years that are now happily spreading and sending out their own suckers. The trees have attractive flowers and foliage with good autumn colour. The trees are also being seriously considered as suitable for planting in Britain for their timber value. Recent research is showing that in Britain their seeds, which normally do not grow successfully in the wild, are remarkably viable when grown using modern nursery techniques.[39] It is possible that warmer summers will allow the species to spread naturally from seed and that the geographical borderlines of reproductive success, which have for example been identified in southeast Denmark, will move northwards in Europe.[40]

Ancient trees

The poet John Clare recorded in his journal for 2 April 1825 a report from his local newspaper, the *Stamford Mercury*, concerning the ancient yew tree at St George's church in Crowhurst, Surrey, where the Lingfield and Crowhurst choir sang 'several select pieces from Handel in the Cavity of a Yew tree'. This remarkable tree was '36 feet in Circumference & is now in a growing state – The hollow was fitted up like a room & sufficiently large to contain the performers.' The news-paper report also noted that when the 'interior of the tree' had been cleared out a few years earlier to make this room 'a 7lb cannon ball was discovered which no doubt had been fired into it; it was cut out from the solid part of the tree.' The peculiar histories and anecdotes associated with such ancient trees have long fascinated antiquaries: the seventeenth-century scholar John Aubrey had noted the huge size of the same 'Ew-Tree, ten Yards in Compass' (illus. 6).[41]

In Britain many of the most famous named trees are ancient oaks. Jacob George Strutt, for example, named 39 individual trees or groups of trees in the 1830 edition of his *Sylva Britannica; or, Portraits of Forest Trees Distinguished for Their Antiquity, Magnitude, or Beauty.* This was a popular edition of the book, which had been published as an expensive folio edition in 1822 for a limited number of sub-scribers. Half of the trees from England included in this collection

6 The Crowhurst Yew, Surrey, 2011.

are oaks. The only other species to have more than one image are elms (*Ulmus* spp.), beeches, chestnuts (*Castanea sativa*) and cedars (*Cedrus* spp.). The names given to the trees, such as the Great Oak of Panshanger, the Great Beech in Windsor Forest and the Great Ash at Woburn, frequently emphasize their size. Others are named after possible historical connections, such as Queen Elizabeth's Oak and Sir Philip Sidney's Oak.

Sherwood Forest in Nottinghamshire is internationally renowned for its ancient oaks. The Royal Forest was established by the twelfth century and in the following century extended over an area around 32 kilometres long and 13 wide on the dry sandstone heaths. In the medieval period the forest was a shifting mosaic of heathland and unenclosed oak and birch woodland with many temporary arable enclosures known as 'brecks'. In the eighteenth and nineteenth centuries much of the heathland was enclosed and converted to parkland and arable fields associated with several large ducal estates. Many plantations of broadleaved and coniferous trees were also established. But some areas of old oaks, which became increasingly linked with medieval legends, survived and were protected by landowners for their aesthetic interest, association with Robin Hood and picturesqueness. One of the most important areas of old oaks was the area known as Birklands and Bilhaugh, just north of Edwinstowe, now part of a National Nature Reserve.

One of the first people to write extensively about the Sherwood ancient oaks was Major Hayman Rooke (1723–1806), a retired army officer.[42] He was a keen archaeologist who excavated a Roman villa at Mansfield Woodhouse and published a meteorological register from 1785 to 1805. He was also an amateur topographical artist who made many landscape drawings. At the end of the eighteenth century he turned his attention to the ancient oaks of Sherwood, publishing *Descriptions and Sketches of Some Remarkable Oaks, in the Park at Welbeck, in the County of Nottingham, a seat of His Grace the Duke of Portland* (1790) and *A Sketch of the Ancient and Present State of*

Sherwood Forest in the County of Nottingham (1799). He was fascinated by the great age, size and form of the ancient oaks and by their royal and other historical associations. He argued, 'Were we, even now, to enter a grove of stately oaks, seven or eight hundred years old, whose spreading branches form a solemn and gloomy umbrage, I think we could not behold them without some degree of veneration.'[43]

As an archaeologist he was keen to ascertain their age and quoted John Evelyn's opinion on tree rings, that 'It is commonly, and very probably asserted, that a tree gains a *new* one every year.'[44] He went on to argue that

> There are now and then opportunities of knowing the ages
> of oaks almost to a certainty. In cutting down some trees
> in Birchland [Birkland] . . . letters have been found cut or
> stamped in the body of the tree, marking the king's reign,
> several of which I have in my possession.

Using these royal ownership stamps, and by counting tree rings, he was able to estimate that one tree, which was sound to its centre, was around three hundred years old when cut down.[45] Rooke was an early adopter of the use of European oak for dendrochronology, which is now of course well established as a most important method of dating timber by archaeologists, art historians and historical ecologists. Dendrochronology has been used at Sherwood to estimate the date of death of a sample of standing dead oaks, the likely longevity of surviving ancient oaks and management implications for the conservation of the dead wood resource.[46]

Nine of Rooke's original drawings were carefully engraved for publication by William Ellis (1747–1810) and Rooke's accompanying text gives full descriptions of the trees, providing valuable insights into his reasons for drawing them and the many ways that he valued large, old trees. One is a portrait of a tree 'remarkable for its height and straightness of trunk' and named 'the Duke's walking-stick'.

Rooke gives detailed measurements and estimates of the volume of timber in the tree and 'doubts whether this admirable tree can be matched by any other in the kingdom'. Another is a view of a 'remarkable ancient oak near the Duchess's flower garden', which is included because of its unusual shape. It was 'called the Seven Sisters, from its having had seven stems or trunks issuing out of one stool in a perpendicular direction'. Rooke made the drawing of this tree and that of the Greendale Oak in 1779, indicating that he had been drawing trees in the area for more than ten years. An unnamed tree is celebrated more 'for the hollowness of its trunk and luxuriancy of its foliage than for size'. The tree trunk 'is hollow from the bottom to the top, like a chimney'. Moreover 'in this tree the game-keeper secretes himself when he shoots the deer, and there are small apertures for his gun.' Rooke adds the detail that 'on the inside is cut 1711; so that this excavation must have been of the same dimensions 78 years ago as it is now, and the tree must have increased but very little in bulk since that time.' He emphasizes that 'Setting aside its hollow trunk, it has every appearance of a young flourishing tree': on one side the bark had been stripped away by a lightning strike, but the tree was 'certainly a remarkable instance of the strength of vegetation, in supporting so large a head on so thin, and, to appearance, almost decayed a trunk'.[47]

The Greendale Oak was already famous for having had a large hole cut through it in 1724 by the 2nd Earl of Oxford following an after-dinner bet that he could drive a coach and six through one of his ancient oaks. Rooke reports that the 'Countess of Oxford, grandmother of the present Duke, had several cabinets made out of the branches, and ornamented with inlaid representations of the oak'. A cabinet remains at Welbeck Abbey today. He provides a general view of the park at Welbeck as it showed that the 'plantations on the hills at the east end of the park are upon a great scale, and truly magnificent'. They included a 'variety of evergreens, forming a pleasing diversity of colours. The trees are chiefly oak and chesnut [sic], with some beach [sic], larch, Weymouth pine [Pinus strobus], and other kinds of firs.'

He emphasizes here the improvement to the landscape brought about by the Duke of Portland's recent patriotic plantations. A contrasting sketch shows a single ancient oak in Clipstone Park to the west of the village of Edwinstowe, 'which the common people call the Parliament-oak, from an idea that a parliament was once held under it'. Rooke could find no proof of this but points out that a Parliament was held at the nearby Clipstone Palace in 1290 and that the tree 'is undoubtedly of very great antiquity'.[48]

The final sketch is of an unnamed tree in Birkland, which he describes as 'a beautiful wood, or rather grove, consisting of above ten thousand old oaks, with birches intermixed', which gave it its name. To the north of the ride cut through the wood 'is a most curious antient oak', which must have been enormous 'before the depredations made by time on its venerable trunk'. He drew a view of the tree from the northeast (illus. 7) showing how the 'trunk, which is wonderfully distorted, plainly appears to have been much larger; and the parts from whence large pieces have fallen off are distinguishable.' He noted that 'the inside was decayed and hollowed out by age.' This meant it was impossible to calculate its age but he thought that 'no one can behold this majestic ruin without pronouncing it to be of very remote antiquity; and might venture to say, that it cannot be much less than a thousand years old.'[49] This tree became known after Rooke's death as the Major Oak to celebrate his enthusiasm for ancient trees.

There have subsequently been many depictions of the Major Oak, including one of 1844 by Henry Dawson (1811–1878) and a large oil painting of 1882 by Andrew MacCallum (1821–1902), both held by Nottingham City Museums and Galleries. There is a wood engraving of the tree exhibited in 1935 by Alec Buckels (1892–1972) in the Scottish National Gallery of Modern Art. But the number of paintings is far outweighed by the huge number of photographs and postcards of the tree produced over the last 150 years, which provide valuable evidence of long-term changes in the tree and its management. Early twentieth-century postcards show how visitors wore away the soil around the

7 William Ellis, after Hayman Rooke,
An Ancient Oak in Birchland Wood, 1790, engraving.

base of the tree and large patches of flowering gorse (*Ulex europaeus*) in the surrounding heathland. Currently the tree is fenced to prevent damage to the soil around it, and there is much less heathland in the immediate surroundings. Contemporary artists use modern scanning techniques to capture formerly secret aspects of the Major Oak. Matt Collishaw's *Albion* (2017) consists of a laser scan of the tree projected to give a spectral image that rotates every half an hour. Caroline Lock (2019) has also taken a laser scan and then cast a bell tuned to the frequency of the tree. Acorns were collected and cast to preserve elements of the 'tree for all time'.[50]

Rooke thought that the Major Oak 'might almost have vied with the celebrated Cowthorpe oak for size' if it had not lost many of its branches. The Cowthorpe oak, near Wetherby, Yorkshire, which no longer exists, was one of several named large old trees that achieved national fame for their size and picturesque appearance.[51] Rooke had not visited the tree himself but knew of it through its mention in Alexander Hunter's edition of John Evelyn's *Sylva* (1776). This edition was the first to contain images of the leaves and fruit of trees, by John Miller.[52] The dimensions of the Cowthorpe oak were 'almost incredible. Within three feet of the surface (1776) it measured sixteen yards in circumference, and close by the ground, twenty-six yards. Its height is about eighty feet, and its principal limb extends sixteen yards from the bole.' He estimated that it would be possible for as many as 'two hundred and fifty-one horses . . . to stand under the shade of this tree'. Hunter pointed out that 'the foliage is extremely thin, so that the anatomy of the antient branches may be distinctly seen in the height of summer.' He felt that this 'venerable tree must once have been the pride of the forest' but that now, quoting the poet Edmund Spenser's *Shepheard's Calendar* (1579),

> His bare boughs are beaten with stormes.
> His top is bald and wasted with wormes.
> His honour decay'd, his branches sere.[53]

Hunter ensured that this tree would be remembered by including a fold-out image of the whole tree, drawn by John Miller as 'A Winter view of the Cawthorpe [*sic*] Oak'. The tree was also sketched in 1816 by J.M.W. Turner.[54] Yet another portrait of the tree was included in Jacob Strutt's *Sylva Britannica* in 1830 (illus. 8). Strutt's text emphasizes the parallels between the oak and the rustic inhabitants of the nearby farm: the ancient branches had 'a peculiar air of rustic vigour, retained even in decay: like some aged peasant, whose toil-worn limbs still give evidence of the strength which enabled him to acquit himself of the labours of his youth.'[55]

Another celebrated oak tree drawn by several artists was the Moccas oak, Herefordshire. Strutt emphasizes the romantic connotations of the tree's position close to the border of England and Wales, which 'is fraught with historical associations, which extend themselves, with pleasing interest, to this ancient "monarch of the wood", among whose boughs the war-cry has often reverberated in former ages'.[56] Through the provision of ship timber, oaks were strongly associated with ancient patrician government and naval power. The tree had earlier been painted by Thomas Hearne (1744–1817), who was invited in the 1780s to Herefordshire by Richard Payne Knight, the connoisseur and author of the didactic poem *The Landscape*, to paint a series of views of his picturesque Downton Gorge. He was then invited to Moccas Court, between Hereford and Hay-on-Wye, by Sir George Cornewall to paint a series of estate views completed about 1788–9, including *The Moccas Oak*. This is a huge, gnarled and hollow pollarded oak tree, one of many in the deer park at Moccas. Although the only use of the timber from the Moccas oak was likely to be firewood, a few years later, in 1798, during the Napoleonic Wars, Hearne's watercolour was etched by Benjamin T. Pouncy and published with

8 Jacob George Strutt, *The Cowthorpe Oak*, 1826, etching.

9 Thomas Hearne, *Specimens of Polyautography:*
Landscape with an Oak Tree, 1803, lithograph.

an accompanying patriotic caption referring to it as 'the source of produce' from whence 'springs the British navy which gives our Island so
honourable a distinction among surrounding nations' (see illus. 12).[57]

The Moccas oaks were celebrated by the diarist Francis Kilvert,
who described a walk on 22 April 1876: 'We came tumbling and

10 Henry William Burgess, *Beech Trees, in the Grounds of St Leonards Hill, Berks, Seat of Field Marshall Earl Harcourt*, 1827, lithograph.

plunging down the steep hillside of Moccas Park, slipping, tearing and sliding through oak and birch and fallow wood of which there seemed to be underfoot an accumulation of several feet.' He saw this as the 'gathering ruin and decay probably of centuries'. He then described 'the vast ruin of the king oak of Moccas Park, hollow and broken but still alive and vigorous in parts and actually pushing out new shoots and branches'. In contemplation he feared

> those grey old men of Moccas, those grey, gnarled,
> low-browed, knock-kneed, bowed, bent, huge, strange,
> long-armed, deformed, hunchbacked, misshapen oak
> men that stand waiting and watching century after
> century, biding God's time with both feet in the grave
> and yet tiring down and seeing out generation after
> generation'. He thought they looked 'as if they had been
> at the beginning and making of the world, and they will
> probably see its end.[58]

Kilvert's diaries were not published until the 1930s, but they became immensely popular and his reference to the Moccas oaks increased interest in their conservation. Moccas Park is now a National Nature Reserve and one of the best surviving areas of wood pasture in Britain.[59] It has several rare species of saproxylic beetles, including the Moccas beetle (*Hypebaeus flavipes*), discovered in 1934, which is only found in Britain in sixteen of the Moccas veteran oaks. For its survival the species needs 'the continuous presence of sufficient old oak trees'.[60]

Thomas Hearne was one of the first artists in Britain to make use of lithography by contributing to the collection *Specimens of Polyautography*, published in 1803. The process had been invented by Alois Senefelder in 1798 to reproduce texts and music. He sold the rights in 1801 and in 1803 a portfolio of twelve lithographs was published by Philipp H. André and James Heath. Thomas Hearne's striking

11 George Childs, *Yew*, 1839, lithograph.

lithograph of an ancient, hollow and collapsing oak tree is one of the earliest tree lithographs ever made (illus. 9). It seems almost anachronistic that a veteran tree is used to exemplify the most modern of print technologies, but such atmospheric tree studies were frequent subjects in early lithographs.

It did not take long for lithography to become one of the most common ways of popularizing tree images. Charles Joseph Hullmandel (1789–1850) met Alois Senefelder in Munich in 1817 and almost immediately decided to publish a set of *Twenty-Four Views of Italy* (1818). He worked closely with a wide range of artists, including James Duffield Harding, Edward Lear, Samuel Prout and James Ward. His influential book *The Art of Drawing on Stone* was first published in 1824 and went through several editions.[61] In 1827 he published Henry William Burgess's *Eidodendron: Views of the General Character and Appearance of Trees Foreign & Indigenous as Connected with Picturesque Scenery*. Burgess (*c.* 1792–1839) had been made landscape painter to King William IV the previous year. As with other published collections there is a mixture of portraits of named, famous trees and those with particularly picturesque characteristics. *Beech Trees, in the Grounds of*

St Leonards Hill, Berks, Seat of Field Marshall Earl Harcourt, for example, shows a pair of huge beech trees in a parkland setting (illus. 10).

By the mid-nineteenth century collections of lithographs of trees had become commonplace. An example is George Childs's *Woodland Sketches: A Series of Characteristic Portraits of Trees, Adapted for Studies for Artists and Amateurs* (1839). In this collection, although specific trees are named in the 'historical, poetical, and miscellaneous descriptions' included in the text, the images provide examples of anonymous trees that were designed to be helpful for artists to use in their own paintings. These are usually placed in characteristic places, such as the lithograph of the yew growing in an old churchyard whose branches embower ancient gravestones (illus. 11).

2

DIVERSE MODERNITY

John Clare's long experience of agricultural labour meant that he
welcomed young plantations of fir trees in the English Midlands
with their 'rich blue-green of summer all the year', which softened
'the roughest tempest almost calm' and offered 'shelter ever still
and warm'. He enthused that 'Winter is almost summer where they
grow.'[1] By contrast, William Wordsworth is well known for his dis-
like of exotic conifers replacing old woodlands in the English Lake
District. His anguish and horror regarding such modernity is indicated by an anecdote that local woodmen told Thomas De Quincey in
the 1830s. A 'reposing band of labourers in the shade' having a midday
break spotted his fury at the replanting of birch wood with larches.
Wordsworth, who did not know that he was being watched, when
he found 'a whole cluster of birch trees grubbed up . . . was heard
pouring out an interrupted litany of comminations and maledictions'.
Then when he saw some larches freshly planted in a line he 'seized
his own hat in a transport of fury, and launched it against the odious
intruders'. De Quincey himself thought larch plantations disfigured
the landscape, having when young 'the formal arrangement of nursery
grounds, until extensive thinnings, as well as storms, had begun to
break this hideous stiffness in the lines and angles'. One had to wait
forty or fifty years before the larch began 'to toss its boughs about
with a wild Alpine grace.'[2]

Across Europe there have been many successful schemes to increase woodland area over the last four hundred years. Many of these consisted of plantations of rapidly growing exotic trees introduced from America and Australia, including species of eucalyptus in Mediterranean countries and Douglas fir and Sitka spruce in northwestern Europe. Some long-established plantations of exotic trees have gained varied cultural, aesthetic and conservation values and raise questions about the relevance of terms such as 'native' and 'exotic'. These modern forms of forestry have often replaced long-standing open areas such as the dry karst landscapes of Dalmatia or the wet moors of Scotland. They also often involve the removal of existing woodland.

Eradicating old trees

Modern forestry has often been strongly associated with the eradication of old trees. In the later nineteenth century huge tracts of Scandinavian coniferous forests were felled, many of which were 'very old Scots pines (250 to more than 600 years old at the time)'. The areas affected included large parts of Sweden, Finland and Russia. The Swedish forest historian Lars Östlund argues that the 'removal of millions of ancient trees across these regions had massive effects on the boreal forest ecosystems, including loss of key ecological features of the forest and was the start of the transformation towards production forest'. The pace of change increased from around 1950 with the popularization of the chainsaw. This allowed the spread of clear-cutting as the main method of felling and harvesting timber. The scale of harvesting was transformed and the profitability of forestry was improved, but there were significant losses of habitat and declines in biodiversity.[3]

Another important development was the extensive use of herbicides from between about 1950 and the start of the 1980s to kill broadleaved trees. In Sweden the popularity of herbicides was encouraged by the sheer scale of felling, which brought about a new management problem: large-scale natural regeneration of birch trees. Bertil Stenlund, a retired forester from Lycksele, Lapland, noted that when hundreds of hectares were clear-felled 'The birch took over. And the poor seedlings of spruce and pine that we planted couldn't compete.' The two main herbicides used in forestry were 2,4-D (dichlorophenoxy-acetic acid) and 2,4,5-T (trichlorophenoxyacetic acid), which cause the 'uncontrolled growth and finally death of deciduous plants and trees'. These two phenoxy acids 'were used together to increase their effectiveness in a mixture known as *Hormoslyr* in Sweden, *Tormona* in central Europe and *Agent Orange* during the Vietnam war in the 1960s'. The forester remembered that Hormoslyr 'came like a rescuing angel'. The herbicides were sprayed on the unwanted broadleaved trees either by hand or by aircraft. They were used to destroy species such as birches, aspen (*Populus tremula*) and goat willow (*Salix caprea*) so that they did not interfere with the growth of Scots pine and Norway spruce, which formed the principal crop.[4]

There was little concern about worker safety in the 1960s. A retired forest worker, Bertil Johansson from Älvsbyn, Norrbotten county, remembered that those who used Hormoslyr had no idea what it was: 'In the forest no one really washed themselves. Hormoslyr was blue as ink. If you cut a potato dumpling to fry, then it was blue.' But concern increased in the 1970s when the dangers posed by the herbicides to humans became known and it was realized that the herbicide could 'penetrate us just as well as it can penetrate the bark of an alder tree'. In the early 1970s there were protests against the use of sprays, for example at Aapua (in Norrbotten), where a villager named Göta Andersson identified several reasons:

first it was about the poison. Most people were against that. But another was that women cleared birches from the forest plantations manually, so they would have lost their jobs. Then there were the berries. People picked them and made money. We sold them and the prices were good.

In addition forest workers became concerned about the impact of the sprays on the woodland. Retired forester Bertil Stenlund remembered 'being horrified that the rowans (*Sorbus aucuparia*) disappeared. In those days there were big rowan trees everywhere in the forest. It was like the forest lit up when you came and saw those beautiful trees and their red berries.' He thought 'the forest became impoverished. I can't say that I could express it at the time but my feeling was that something awful has happened here.' One of the forest activists, Magnus Sjögren, decided to protect some of the trees he had been employed to kill:

> There were large aspen trees . . . 80 cm in diameter, which we were to kill. I thought no! Damn it! So then I made the notches with the axe, the others will see me since we work together, so I faked it. I made the little notches but I did not pour in the herbicide. So they could live on.

The protests were effective and forest companies began to reduce the use of these herbicides in the later 1970s. Their use was banned in 1984. Overall, it is estimated that around 700,000 hectares of Swedish forest was sprayed between 1948 and 1984; the long-term ecological impact of the use of such herbicides remains to be determined.[5]

In the UK the use of the herbicides 2,4,5-T and 2,4-D continued to be allowed in forestry until 2009, more than twenty years after the Swedish ban in 1984. The Forestry Commission's guidance on the chemical control of woodland weeds published in 1975 remained enthusiastic about their use, pointing out that, after many tests of different chemicals on woody broadleaved weeds, none were 'so generally effective or so safe to the crop, user and environment as 2,4-D, 2,4,5-T or ammonium sulphamate'. There was concern about the 'unsightly' appearance of 'large standing dead trees', especially if they were in an 'isolated or prominent position', and it was recommended that large trees adjoining roads and footpaths should not be sprayed as landowners 'may be liable for injury to passers-by or damage to their property from falling branches'. But otherwise, woodland owners were encouraged to use herbicides to kill a wide range of broad-leaved 'weed' trees including beech, oak, hornbeam, lime, field maple, elm and hazel.[6] Another way of killing unwanted broadleaved trees that remained popular until the 1970s was to remove a ring of bark from the main stem. This is known as ring-barking and was commonly used. I remember as a student attending a meeting with the Forestry Commission to try and encourage them to stop the ring-barking of old oak trees in the Forest of Dean as late as 1976. In the same period many ancient oaks were ring-barked by the Forestry Commission in Croft Wood, north Herefordshire. At Balgownie Wood, Fife, there still remain many old oak coppice stools which had been ring-barked as preparation for the first planting of conifers by the Forestry Commission in the late 1930s.[7]

Foresters have, over the last thirty years or so, increasingly recognized the importance of retaining a proportion of large old trees for cultural reasons and nature conservation. Many countries, such as Poland, have legally protected trees that are listed and mapped. In the Kampinos Natural Forest on the outskirts of Warsaw, for example, 69 trees are classed as nature monuments, of which the majority are oaks (56) with six Scots pines and five small-leaved limes (*Tilia cordata*). In the UK volunteers and professional arboriculturist members of the Ancient Tree Forum map ancient trees, judge their condition and lobby for their protection. In addition new methods are being developed to identify veteran and ancient trees within woodland and growing in open countryside.[8]

Eucalyptomania

Some of the fastest-growing trees are various species of eucalyptus and these are increasingly planted in many countries. Worldwide it is estimated that there are over 20 million hectares of eucalyptus plantations that are cropped at an industrial scale to provide construction timber, wood pulp for paper, and domestically for firewood and charcoal.[9] Such plantations are also seen as having great potential for carbon sequestration: in Cornwall, for example, Hugh and Tina Davis have planted 16 hectares of eucalyptus since 2014 on their family farm near Wadebridge for firewood production, and point out that eucalyptus trees produce '40 cubic metres per acre per annum in growth compared with 18 cubic metres for Sitka spruce' and that they 'absorb about 20 tonnes of carbon dioxide per acre compared with 4.5 tonnes for oak'.[10] But there is another side to these trees. They can cause soil degradation and are increasingly associated with forest fires, causing widespread destruction and death. One of the most horrifying examples occurred on 17 June 2017 when intense forest fires spread rapidly through woodlands in central Portugal near the small town of Pedrógão Grande. The fires killed at least 66 people, many of whom were trapped in their cars as they tried to escape the flames. The fires took hold and spread through eucalyptus woodland, which is now characteristic of many parts of Portugal. Indeed, eucalyptus trees cover around 900,000 hectares, almost a quarter of the total area of Portuguese woodland. For many people, both in Portugal and internationally, the eucalyptus has been designated a 'killer tree' whose flammability has become notorious.[11] How have these trees become such an important element of the forest economy of Portugal and other countries?

There are around eight hundred species of eucalypt and almost all of these are native to Australia. Today they cover just over 100 million hectares, more than three-quarters of Australia's native forest. Most eucalypts are evergreen and they have leaves rich in inflammable oils. The trees have evolved to survive and recover from fires. Traditionally, indigenous Australians made use of all parts of eucalypts: the leaves for medicinal purposes and the wood and bark for tools, weapons and making canoes. Some of the most striking evidence of such use is provided by the surviving 'canoe trees', usually red gums (*Eucalyptus camaldulensis*), which retain distinctive scars showing where bark was removed to make canoes and other objects. A photograph taken in the early twentieth century of large old eucalyptus trees in the Murray River region clearly shows the marks in the bark where canoes have been cut (illus. 13). Such trees are found across south and east Australia but 'are one of the least understood elements of Aboriginal heritage' and research is being undertaken to discover more about 'their age, origin and interpretation'. Early photographs such as this one by Harry Godson are an important resource for this type of research.[12]

Following the arrival of Captain James Cook's *Endeavour*, with the keen botanist Joseph Banks, at Botany Bay on 29 April 1770 and the subsequent claiming and naming of New South Wales on 22 August 1770, there was great enthusiasm in Europe for Australian plants and animals. It is thought that the first eucalyptus planted in Europe was in 1774 at Kew Gardens, London. This was *Eucalyptus obliqua*, from seed collected by Captain Tobias Furneaux, who commanded *Adventure* during Cook's second Pacific voyage. The seeds were collected from Adventure Bay, Bruny Island, Tasmania, in March 1773.[13]

It was soon realized that some species of eucalyptus could be grown outside glasshouses in Britain. John Claudius Loudon (1783–1843), writing in the 1830s, recognized over a hundred species or varieties, all of which were 'timber trees, growing to a great height, and natives of New Holland and Van Diemen's Land'. He thought that those 'belonging to the latter country appear to be decidedly half hardy in the neighbourhood of London'. He was excited by the idea of planting many trees together in southern England to form an 'entire wood' when the trees 'would protect one another; and, if they did not attain the size of timber trees, would, at least, form a dense Australian copse.' He hoped that landowners would 'encourage their gardeners to plant out these, and other Australian trees, in dry sheltered places in their shrubberies and woods', noting that plants were cheap and 'seeds are very frequently imported, and might be sent home in the greatest abundance if there were a demand for them.'[14]

However, by the end of the nineteenth century this bubble of British enthusiasm had at least partially been burst. The leading arboriculturalists Henry Elwes (1846–1922) and Augustine Henry (1857–1930) considered that no genus of trees 'has been more persistently tried in various parts of the country; and yet when we come to record the small number of trees that have endured our climate for more than a few years, it must be acknowledged that none has proved more disappointing.' Generally in Britain the trees 'are short lived, and die off suddenly after an inclement season, or blow down, when they become tall enough to be exposed to gales'. They conclude that none of the species were 'likely to produce timber of any commercial value in Great Britain'. They did, however, report on one or two well-established stands:

> The most remarkable plantation of Eucalyptus we know
> of in England was made by the late John Bateman of
> Brightlingsea Hall, Essex, who raised seeds of *Eucalyptus*
> *gunnii*, which were sent him from southern Argentina
> in 1887, by Mr Shennan, who had naturalised there the
> Tasmanian *E. Gunnii* . . . some of the trees have now
> attained a height of 40 to 50 ft, with a girth of 3 or 4 ft.

These trees were regenerating rapidly. On 9 December 1906 'some were in full flower.' They 'ripened seed every year'. They seemed to like wet areas and 'a great number had germinated in the gutter of the coach-house.' This globetrotting eucalyptus was clearly making itself at home in coastal Essex.[15]

Eucalyptus trees were widely planted in Mediterranean Europe. It is thought that the first ones planted in the open air were at the English-style garden of the Queen of Naples at the Royal Palace of Caserta in 1792. The seeds were probably sent to Naples from Kew by Joseph Banks via Sir William Hamilton. Eucalyptus trees were grown in botanic gardens and parks throughout Italy, Spain and Portugal in the early years of the nineteenth century. They were valued for their novelty and beauty and their rapid growth meant that they soon became grown for timber. In 1852 in Portugal, for example, the Marquis de Massarelos made plantations at his Quinta da Formiga estate of at least four species of eucalyptus.[16] The editor of *La Belgique Horticole* claimed in 1871 that 'a real Eucalyptomania now reigns throughout the temperate world . . . It is a precious tree. Its wood is hard and magnificent.'[17] But perhaps the most important reason for the popularity of eucalyptus in the later nineteenth century was its perceived ability to improve the healthiness of places where it was planted. Its rapid growth meant that it could be used to dry marshy areas associated with malaria.

The key figure in the popularization of the potential medicinal benefits of eucalyptus was Sir Ferdinand Jakob Heinrich von Mueller (1825–1896). He studied pharmacy at the University of Kiel in 1845–7 and gained his doctorate on the flora of Schleswig. He moved to Australia for health reasons and soon started to survey the South Australian flora, publishing his findings in international journals and local newspapers.[18] He encouraged the commercial distillation of eucalyptus oil and was 'responsible for exporting eucalyptus seeds to California, India, Algeria, Hong Kong and elsewhere, advocating their planting as a measure to combat malaria'. Between 1879 and 1884 he published his influential *Eucalyptographia: A Descriptive Atlas of the Eucalypts of Australia and the Adjoining Islands*. Mueller was a keen advocate of the health benefits of planting eucalyptus. In 1880 he told the Marchese di Toverena, the Italian Consul General in Australia, that 'malaria could be restrained if eucalpts were extensively cultivated in La Campagne.'[19] More generally, his publications helped persuade the pioneer American conservationist George Perkins Marsh (1801–1882) of the advantages of tree plantations. In the 1864 edition of *Man and Nature*, Marsh had argued that plantations of trees 'could not compete economically with natural forest: therefore it was altogether better to preserve America's forests so as not to make the mistake of the European experience.' But in later editions he used examples of successful eucalyptus plantations in Italy and California as evidence for the advantages of acclimatization and afforestation. Marsh noted that in California 'the child is perhaps now born who will see the tallest sequoia overlapped by the new vegetable immigrant from Australia.'[20]

In Portugal extensive plantations of eucalyptus were made from the 1960s onwards, mainly to produce short-fibre paper pulp.[21] By the mid-1980s around 300,000 hectares of *Eucalyptus globulus* had been planted on former vineyards, olive groves and pastures that were deemed unproductive and uneconomic. Eucalyptus plantations (4 person days per ha) required substantially less labour than vineyards (128 person days per ha) or olive groves (199 person days per ha). A review of the advantages and disadvantages of eucalyptus published in 1986 noted that 'Eucalyptus can be counted on to produce at least twice as much pulp wood as pine does.' Moreover, rapid coppice regrowth meant that 'after a cutting operation, two or three more harvests from each stump are possible.' Several environmental risks were identified, including the threat of soil erosion from deep ploughing and the possible loss of soil fertility. It was also noted that there were likely to be adverse effects on nature conservation, although this was strangely discounted because, 'Traditionally, protection of the natural fauna and flora has had no great status in the countries of the Mediterranean basin.' The review recommended that the adverse effects of the eucalyptus monocultures on the landscape could be mitigated by mixed plantings along the forest edges, which would help to break up 'the quadrangular pattern both within and between stands.'[22]

Perhaps the most startling aspect of the review is the way in which it downplays the fire risk. It accepts that 'Fires are a topic worthy of special mention as eucalyptus species may sometimes be easily combustible. Bark and litter collect on the ground, and bark strips hang from the trunks of trees older than eight years, which presents a risk in areas with pronounced drought periods.' However, it then argues that 'there is evidence that pine, with its looser and more combustible litter and its high phenol content, is more likely to burn than eucalyptus.' It goes on to note extraordinarily that 'Planting eucalyptus globulus does not prevent fires, but seen against the background of the investments made in these plantations, it is probable that a more active fire-prevention and guarding of the forest will result.' One of the main conclusions of the review is that 'planting *Eucalyptus globulus* in Portugal for the production of pulp wood with short, 15-year rotation periods is not likely to pose any tangible threat to the long-term productive capacity of the soil.' Moreover, the risk of erosion 'will be balanced by a better supply of organic elements and a more long-term stability encouraged by improved fire prevention.'[23]

Today it seems incredible that the increased risk of fire was not taken more seriously in the 1980s. It is true that there were some protests against eucalyptus, including one at Veigo do Lila in Valpaços where 200 hectares of recently planted eucalyptus were uprooted in March 1989 by a crowd of local people shouting 'Yes to olive trees yes, No to eucalyptus.' This particular protest was successful.[24] But in most areas eucalyptus continued to be planted, and although they are recognized as causing widespread and dangerous fires, 'ever more forest owners have switched to eucalyptus. Hoping that a shorter production cycle might allow them to recoup their losses faster and harvest their trees before the next fire erupts.' In addition, eucalyptus trees are gaining popularity because of their ability to assist in carbon sequestration.[25]

Pines on karst

Afforestation has long been seen as a way of improving areas with impoverished, limestone soils such as the karst landscapes of Dalmatia. After the collapse of the Venetian Republic in 1797 Dalmatia had short periods of French and Italian control before becoming the Kingdom of Dalmatia integrated within the Austrian Empire (1815–66) and Austro-Hungarian Empire (1867–1918). The Academy of Forestry at Mariabrunn monastery, Vienna, founded in 1813, was a leading centre for the spread of modern forestry ideas through the empire, including coniferous afforestation. Vast areas of Dalmatia's arid, karst landscapes were grazed by flocks and herds of sheep and goats, while along the coast and on numerous islands there was intensive terraced cultivation of vines and olives, together with fishing. Traditional coastal woodland included Mediterranean species such as holm oak (*Quercus ilex*), mastic tree (*Pistacia lentiscus*), terebinth (*Pistacia terebinthus*) and mock privet (*Phillyrea latifolia*). Pine woodlands, mostly Aleppo pine (*Pinus halepensis*), were found only along the coast and islands in the far south. In the hinterland these trees were replaced by species such as manna ash (*Fraxinus ornus*), downy oak (*Quercus pubescens*) and different hornbeams (*Carpinus* spp.) and junipers (*Juniperus* spp.). Such woodland areas were rare but a crucial local resource for grazing, browsing, leaf fodder and firewood, and were usually managed as coppice with some pollards.[26]

Nineteenth-century Austrian and Dalmatian foresters were keen to emphasize that the original rich woods growing on the karst had been destroyed by the Venetians. They argued that karst, until a few hundred years ago, was covered with lush, high forests and that this vegetation had been destroyed through deliberate cutting. The Venetians, who were 'greedy and soulless merchants, lustful only for riches and wealth', had felled most Dalmatian forests and left 'devastated and barren karst' as their legacy and monument. Another reason for the 'spreading of karst' was local people whose 'uncontrollable use of pasture' was 'the demon devastating the hillsides'. Joseph Wessely (1814–1898), head of the Forestry Academy in Vienna, blamed the lack of natural regeneration of trees in Dalmatia on the poverty of the local inhabitants and their reliance on goats and sheep for their livelihoods (illus. 14). This is why, he argued, 'wherever we look or reach in that horrible edge of our otherwise advanced Monarchy, everything is desert and bare.' Another forester observed local people cutting and digging out stumps from communal areas in order to sell them for firewood at coastal markets.[27]

Although some commentators, making use of historical maps, documents and archaeological findings, argued that karst landscapes had been in existence long before the Venetians arrived, the move towards coniferous afforestation was irresistible. One of the main goals of Austrian foresters was the 'reforestation of karst: through the growth of tall trees to create ... woodlands, pastures and the thick layer of humus which had allegedly existed before it was made barren'. The idea that reforestation would lead to the replenishment of the soil and economic recovery through cultivation of newly created productive areas had caught hold.[28]

Early attempts to afforest the hinterland of Trieste demonstrated that the Austrian or black pine (*Pinus nigra*) succeeded better than broadleaved trees, while the Aleppo pine became the most planted species in Dalmatia.[29] Some exotic broadleaved trees were planted, including the tree of heaven (*Ailanthus altissima*), which some argued was resistant to goat browsing, and eucalyptus for some marshy areas. But local species such as the holm oak were ignored. The dominant idea was to create high forests, largely of pine. This ignored the importance of sheep and goats in the pastoral economy, which Wessely thought had to be considered if afforestation was to be successful. He favoured the establishment of some woods specifically designed for browsing animals. These would be professionally managed areas of broadleaved trees providing fodder for livestock, including ash, hornbeam, varieties of oak, beech, mulberry and cherry. Such woodlands for browsing, he believed, would improve the pastoral economy and reduce the pressure on meadows and other woodland areas, allowing their recovery.[30]

This suggestion was largely ignored and consequently reforestation was very unpopular with local people. Foresters were well aware of their unpopularity. In 1889 Dalmatian municipal foresters warned the Dalmatian Parliament about the dire conditions in which they worked. They explained that local people saw reforestation as an unjustified theft of land and that they took revenge by attacking foresters and destroying newly planted trees. In the coastal city of Šibenik the authorities suggested that the focus should be on reforestation schemes in places where opposition would be low. But this did not always work. For example, in the district inland of Šibenik, villagers threatened forest workers in 1904: a forester in charge claimed that the local village chief and several armed goat keepers were among the protesters who refused to allow tree planting to continue. The district authorities sent an armed escort of four soldiers to protect the workers and the planting proceeded. Records confirm that incidents like this were not unusual. A few years earlier, in 1899, the municipal forester and workers at Bilo were attacked by twenty villagers from the nearby villages of Tribešić and Pod Greben. It was reported that the villagers destroyed all the new boundary markers for the area to be afforested and 'shouted and swore they would not allow reforestation anywhere'. In 1907 the foresters and the district authorities were negotiating with various village councils to try and agree new places for reforestation. Since reforestation was dependent on the will of village councils, regulations were implemented very differently from village to village. In practice, Wessely's point that there could be no reforestation without the consent of the people was shown to have been correct.[31]

Reforestation was brought to a dramatic halt by the First World War, which also, of course, destroyed the Austro-Hungarian Empire. However, in terms of the establishment of new pine woodland, Dalmatian reforestation at the turn of the nineteenth century could be deemed a success. Indeed, most pine woodlands that were recorded in the mid-twentieth century in Šibenik district had been established in the pre-war years. Forestry statistics from 1957 indicate that 86 per cent of pine woodland was between 41 and 60 years old and hence had been planted between 1898 and 1916. But what about remnants of the broadleaved woodlands? In 2019 the geographer Ivan Tekić took me to a remote site near the village of Vrsno, 22 kilometres southeast of Šibenik, to see if we could discover any surviving old broadleaved trees. Within a walled enclosure we found eight very large oak pollards. The largest had a girth of over 5 metres, with four surviving old pollard stems estimated at seventy years (illus. 15).

These old trees were all close to terrace walls within the enclosure. The main pollard branches were around sixty years old. It is likely that these pollards used to be cut for leaf fodder and there was evidence provided by gnarled old scars of frequent cutting, which stopped in the 1960s. Between the ancient oaks we found many naturally regenerating trees and shrubs including oak, juniper, ash, oriental hornbeam (*Carpinus orientalis*) and butcher's broom (*Ruscus aculeatus*). There were also some young pines of around fifteen years old. A cadastral map of 1825 indicates that there were small arable fields within the enclosure, and an aerial photograph of 1968 indicates that the area was not overgrown with trees. It seems likely, therefore, that the young growth used for leaf fodder was doubly protected from sheep and goats, both by the walled enclosure and by being grown as pollards. The system was largely abandoned by the 1960s, and now there is rapid natural regeneration of shrubs and trees. The ancient oaks survive as memorials to a forgotten way of life.

Sprucing moors

A tree that is now the most frequently planted and sometimes hated species in the UK made a quiet entry to the British arboricultural world. The Sitka spruce, once also known as the Menzies spruce, was 'discovered in Puget Sound in May 1792' by Archibald Menzies (1754–1842), who was naturalist and surgeon 'on the Discovery, under Captain George Vancouver. The party was to explore and chart the coast of north-west America.' It was later found and described by David Douglas in 1825 in the same area and 'introduced to cultivation' by him in 1831. Loudon wrote in 1844 that 'Only a few plants . . . were raised in the Horticultural Society's Garden in the year 1832; so that the species is at present extremely rare in this country' but 'it is . . . much more plentiful in Scotland.' The name was internationally agreed as *Picea sitchensis*, named after the town of Sitka in southeast Alaska, in 1855.[32]

The natural range of the species is throughout the coastal region of Alaska and British Columbia and down through Washington and Oregon to northern California. It became well known as a moisture-loving species and foresters were impressed by its speed of growth, which was 'very rapid, the leading shoots of young trees on Puget Sound being often 3 to 4 feet long'. They were also impressed with their great age, Henry Elwes noting that 'A tree measured by [John] Muir at Wrangel, Alaska, was no less than 764 years old, with a trunk 5 feet in diameter, and this, I think, is the greatest age to which any recorded spruce has attained.' The Oregon Botanical Association, formed in Edinburgh by keen Scottish arboriculturists to help introduce conifers from the Pacific coast, 'were fortunate in procuring a large quantity of seed, from which the pineta of Scotland and England have been stocked, and it has now [1904] become a common tree'. Elwes, writing in 1904, noted that it 'loves a wet climate, it loves a wet soil even more' and 'in some parts of Scotland it is now being planted experimentally as a forest tree.'[33] Writing on English estate forestry in the same year, Arthur Forbes, who was soon to move to the Irish Forestry Department, becoming the first Director of Forestry in the Irish Republic, agreed that 'The Menzies Spruce' made 'its best growth on damp but porous soils of a peaty nature'. He pointed out that 'it must be grown clean to be of any commercial value', that is, the trees must grow up close together to minimize the number of side branches. But until seedlings could be produced 'at a cheaper rate it is not likely to be planted close enough on a large scale to effect this'.[34]

In the years up to the First World War many private foresters experimented with growing Sitka spruce. At Durris in Aberdeenshire, for example, the forester John Crozier reported that there was a 15-acre plantation of Sitka spruce made in 1879 'at an altitude of 700 to 800 feet'. His experience, based on 'what I have seen here and elsewhere', was that the Sitka spruce 'will stand a greater degree of moisture than any other conifer I know. The plantation is altogether in a very healthy state.'[35] In 1910 he summarized his knowledge and experience of growing the tree in an article for the Royal Scottish Arboricultural Society. He felt that of the many tree species successfully introduced from western North America none had greater potential than Sitka. Crozier thought that it was 'a matter or regret' that, although the tree had been known for nearly eighty years and 'appreciated in pineta and pleasure grounds, on account of its ornamental value', very little 'has been done in a practical way, to ascertain its commercial value'. He agreed with Forbes that close planting was essential, pointing out that 'Sitka spruce responds readily to cultural management, and while in isolation the trees usually carry the maximum proportion of branch to bole, in close canopy stem-cleaning can be brought about by the minimum of side shade.' Its timber was 'not liable to warp, is suscep-tible of a beautiful polish, and when varnished mellows down, in time, to a rich yellow brown'. Overall he argued that Sitka spruce was to be recommended for afforesting exposed sites, particularly in the Scottish highlands. He thought that 'grown in high-forest, it has great productive capacity and yields a high quality of timber' and in addi-tion 'enjoys practical immunity from attack by insect and by fungoid pests'. The only disadvantage was 'the present high price of plants precludes the possibility of pure planting' but he hoped that with increased demand the price would become similar to that of 'ordinary commercial conifers'.[36]

The establishment of the Forestry Commission in 1919 boosted the demand for afforestation of exposed, damp moors, which were otherwise seen as of little agricultural value. Forestry Commission experiments at Beddgelert Forest in north Wales were set up to study the influence of exposure on five species that earlier experiments by private landowners indicated had great potential for upland afforestation: Sitka spruce, Norway spruce, Japanese larch (*Larix kaempferi*), European larch (*Larix decidua*) and lodgepole pine (*Pinus contorta*). The species were planted in 25 plots with different treatments and mixtures. It was 'the first field experiment anywhere in the world' to use one of the statistician Sir Ronald Fisher's Latin square 'randomised designs' to allow comparison. The experiment showed conclusively that 'whether planted pure or in mixture, the volume production of Sitka spruce was far superior to either Japanese larch or lodgepole pine.'[37]

After the Second World War, with the expansion of afforestation by the Forestry Commission, Sitka soon became the most commonly planted tree in the uplands. The Forestry Commission stated in 1970 that 'for many years' it had 'planted more Sitka spruce than any other individual kind of tree', especially on 'peaty hills and moors ... Where much poor grazing land has become available for afforestation'. The tree was valued because 'it yields a greater volume of timber, in a given time, than any other tree. Further it grows upright despite severe exposure, even to salt-laden winds blowing straight in from the sea.' The timber was used for pit props, fencing poles, boxes, joists, rafters, flooring, chipboard and paper pulp.[38]

The leading forester Alan Mitchell described Sitka spruce in 1972 as 'A tree of immense vigour over a long period'. Once seedlings were established 'on damp sites and particularly regions with damp air, like Argyllshire, shoots become regularly three, four and occasionally five feet long. Some trees are still growing annual shoots of three feet when they are 120 feet tall.' He thought 'more intensive treatment of sites before planting is justified economically' because it enabled 'Sitka spruce to establish itself rapidly' in places where growth 'would otherwise be too slow'.[39] The type of treatment Mitchell referred to here was the ploughing of peat and the application of fertilizers. By the 1960s 'Giant ploughs pulled by powerful tractors make it possible to plant successfully on sour, peaty soil which not so long ago would have been regarded as useless.' The Sitka seedlings were 'planted in the over-turned ridge which is thrown up by the ploughs as they make the deep drains required to dry out the moorland'. The young trees grew well in the relatively dry, peaty soils, especially as phosphate fertilizers were usually applied.[40] Between 1957 and 1988, 88 per cent of all afforestation by the Forestry Commission consisted of Sitka spruce (illus 16). In Ireland in 1948 Sitka spruce formed 18 per cent of the area of Irish state forests; by 1970 this proportion had increased dramatically to 70 per cent.[41]

The expansion of coniferous plantations in the uplands had been contentious since the mid-nineteenth century. In the 1930s there was an outcry over the planned afforestation of parts of the Lake District and commercial forestry was often viewed as antithetical to pleasing landscapes. Ronald Knox, the biblical scholar and detective novelist, wrote, 'and the woods grew thicker, their gloom more profound, to show that they were cultivated not for amenity or for sport, but as marketable timber.'[42] After the war critics talked of the 'great dark angular slabs of the Forestry Commission's plantations spreading over the hills like a disease'. Such new plantations were not only disliked because they appeared foreign, gloomy and alien; the fencing and ploughing also restricted public access. One of the key pressure groups opposed to the planting schemes was the Ramblers' Association, which in *Afforestation: The Case against Expansion* (1980) noted, 'At the time of writing, furious battles are being fought against afforestation pro-posals in the Brecon Beacons and Snowdonia.' The new 'man-made' forests 'are dominated by a small number of tree species; notably the Sitka spruce, which alone occupies 32 per cent of the total productive conifer area of Great Britain.'[43]

The move to planting on peat brought about the most critical, sustained and effective attack on Sitka spruce. It stemmed from the expansion of planting onto the blanket bogs of Caithness and Sutherland in the far north of Scotland. To many people these were seen as bleak areas that produced little of economic value. But ecol-ogists and ornithologists identified the extensive peat bogs as 'natural

Post-glacial climax vegetation'. A 1987 report by the Nature Conservancy Council was heavily critical of the afforestation, emphasizing that the bogs supported breeding populations of birds such as golden plover, dunlin, greenshank and breeding red-throated and black-throated divers, and were a globally significant 'primaeval ecosystem'. The deep ploughing and draining essential for successful afforestation destroyed this habitat and the report pulled no punches in asserting that 'coniferous afforestation is destroying these peatlands' and that the threats from state and private forestry were on a scale that 'would have seemed inconceivable' twenty years earlier. The campaign against such upland planting on peat was successful and the Flow Country is now recognized as 'the most intact and extensive blanket bog system in the world'. It became the first peatland UNESCO World Heritage Site in 2024.[44] The campaign also helped to halt the planting of conifers more generally on 'unimproved land in the English uplands', with Nicholas Ridley, Secretary of State for the Environment, announcing in 1988 that 'approval should not normally be given in these areas for new planting which consisted predominantly of conifers.'[45]

Following the restriction of afforestation on valuable moorland sites, there are signs that attitudes to Sitka have started to change from the low point reached at the end of the twentieth century. Current research on the carbon sequestration potential of Sitka is indicating a formerly unconsidered benefit. Forest Research is comparing the carbon balances of two forest stands: an upland Sitka spruce plantation in Harwood Forest, Northumberland, and a lowland mainly deciduous oak plantation at Alice Holt Forest in Hampshire. Methods including eddy covariance flux towers, used 'to determine the rate of exchange of trace gases and energy between ecosystems such as forest . . . and the atmosphere', and 'soil gas exchange chambers', have 'continually measured, hourly and daily, the uptake and release of CO_2'. The measurements started in 2013 at Harwood and 1999 at Alice Holt. These measurements have been used together with measurements of the trees themselves, including height, girth and crown size, to establish the different carbon balances between the coniferous and broadleaved stands.

Overall it was found that 'total carbon stocks, including tree stems, roots, litter and soils' were higher in the dense Sitka forest 'on a peaty-gley soil' compared to the lowland oak forest. Moreover, the 'average annual net ecosystem CO_2 uptake was higher at the spruce forest' as the dense evergreen canopy 'remained photosynthetically active all year round'.[46] The ecologist Ruth Tittensor points out that the cultural histories of Sitka spruce forest are inherently interesting and valuable. They 'have social histories, heritage value, ecological features and futures equal to their alternatives in Britain and Ireland'. As time passes they are 'gradually becoming acceptable in upland landscapes' and she hopes that 'some naturally regenerated Sitka spruce tracts in high-rainfall uplands will progress towards new-style temperate rainforests without expectations (or interferences) on our part as to their structure and processes.'[47] But protests against Sitka continue. In August 2023 protesters at County Leitrim, Ireland, 'were joined by climate activists from across the country . . . as they pulled up hundreds of Sitka spruce saplings' planted on land managed by the government-owned forestry company Coillte. A spokesman from the Save Leitrim campaign argued, 'The bog from which people pulled up the Sitka saplings today would sequester and store more carbon than those Sitka spruce trees ever would.'[48]

Rewilding

The idea of rewilding different landscapes has caught the public imagination over the last twenty years or so. Ecologists describe rewilding as 'a novel and rapidly developing concept in ecosystem management, representing a transformative approach to conserving biodiversity.'[49] The term 'wildwood' was introduced by Kenneth Grahame in his enormously popular *The Wind in the Willows* (1908), based on childhood recollections of Quarry Wood, near Cookham in Berkshire. He drew on ideas of dense, dark woodland full of danger central to children's tales collected by the Brothers Grimm in the nineteenth century and his wildwood was dominated by weasels and stoats who threatened Toad, Ratty and Badger. Foresters and ecologists such as Herbert Edlin (1947) and Oliver Rackham (1980) used the term wildwood to describe some sort of natural woodland before it had been strongly influenced by humans.

Rewilding is closely associated with changes in land management designed to encourage certain species, most frequently animals, to spread and live with limited human intervention. The ideas behind rewilding were initially developed independently in contrasting landscapes and habitats. One of the most important origin sites was Yellowstone National Park, where the reintroduction of wolves in the 1990s led to predation of elk and changes in vegetation. In Europe a key rewilding site is Oostvaardersplassen on the entirely artificial South Flevoland polder northeast of Amsterdam. In Britain a particularly influential experimental site is on the 1,400-hectare Knepp landed estate in Sussex, where rewilding since 2001 has allowed shrubs and trees to regenerate naturally over former agricultural land to form a mosaic of habitats. At Ennerdale in the Lake District, a rewilding scheme covering 4,750 hectares was set up in 2003 to encourage natural processes, including woodland regeneration. These significant European schemes are at quite a small scale, but they have been very influential in popularizing the idea of rewilding.

The Dutch conservationist and biologist Frans Vera stimulated interest in the relationship between grazing animals, trees and woodland. His book *Grazing Ecology and Forest History*, published in 2000, focused attention on grazed, open woodlands and the extent to which these were more typical of much European woodland after the last ice age than dense forest. Vera proposed that large populations of mammals such as elk, aurochsen, beaver and boar helped to create and maintain a substantially open landscape over tracts of northwest Europe. There would have been a shifting mosaic of open areas, patches of naturally regenerating trees, groves of mature trees and scattered, single open-grown trees. The old idea of dense forest, Vera argued, had been derived from an overemphasis on the tree and shrub component in early palynological studies', the misinterpretation of early literary references to forests and woods, and the exclusion of the impact of grazing from many studies of forest succession. Vera's work influenced a generation of conservationists who were developing management plans for different types of wood pasture across Europe, what Oliver Rackham had termed 'savanna in Europe'.[50]

Vera developed his ideas at Oostvaardersplassen nature reserve on the South Flevoland polder reclaimed from Lake IJssel in the 1960s. The reserve consists of 6,000 hectares of marsh and wet and dry grasslands with shrubs and trees and open water, and is an outdoor laboratory for the observation and analysis of the developing relationships between geese, grazing animals such as Heck cattle and Konik ponies, and vegetation. Vera stresses the need to 'develop large, natural-functioning areas where natural processes get the chance to evolve'. He argues that we shall 'have to learn to co-exist with animals living in a truly wild existence, periodically losing condition, and a number of animals dying off as a result of lack of food. If we are unable to do this, we run the risk of making the presence of unfettered nature impossible.'[51] Some conservationists worry about the animal welfare associated with such 'unfettered' rewilding, while others 'fear the loss of habitats for rare species' and 'are anxious at the demise of cultural landscapes'.[52] Others stress that the context of starting rewilding with 'the tabula rasa of a polder reclaimed from the sea in the Netherlands' is very different from 'starting with a wood-pasture in Romania, a large tract of moorland in the Hebrides or, indeed, a mountainside in the Pyrenees'. The key difference is that when one begins rewilding a site with existing conservation value, the likelihood of losing characteristic habitats, species and conservation value through the abandonment of traditional land management practices can outweigh the potential benefits from rewilding.[53]

Irrespective of these concerns, the idea of rewilding has developed a lively potency, and it is being put into practice in several parts of Europe. George Monbiot's 2013 book *Feral: Searching for Enchantment on the Frontiers of Rewilding*, which draws on Vera's research, stimulated widespread and intense public debate in Britain about the balance between farming, nature conservation and recreation. He brought the public benefits of subsidies to hill sheep farming under scrutiny and explored the potential benefits of allowing natural regeneration of woodland over open moors and fells. At Ennerdale in the Lake District the three bodies that own the valley (Forestry Commission, National Trust and United Utilities) established the Wild Ennerdale Project in 2003 'to allow the evolution of Ennerdale as a wild valley for the benefit of people, relying more on natural processes to shape its landscape and ecology'. The purpose here is not to recreate an historical past state but for natural processes to lead changes in the almost 5,000 hectares of upland grazing land and broadleaved and coniferous woodland. But large-scale interventions may still have to take place. For example, in 2021 *Phytophthora ramorum* disease was identified in 60 hectares of larch plantations in the valley. Forestry England felled all the infected trees by 2023 to stop the spread of the disease to other trees and bilberry (*Vaccinium myrtillus*). The trees will be replaced by natural regeneration and planting of native tree species.[54] In 2011 Rewilding Europe, a Dutch foundation, was established to encourage the rewilding of diverse landscapes across Europe. By 2023 they had ten projects running in the Iberian highlands, the Rhodope mountains, Bulgaria, the Coa Valley, Portugal, the southern Carpathians, the central Apennines, Swedish Lapland, the Velebit mountains, Croatia (see illus. 14), the Danube delta, the Oder delta and the Affric highlands, Scotland.[55]

Rewilding does not have to be part of a consciously and carefully designed set of experimental reserves. In Britain there are many examples of formerly grazed common land and heathland that have become wooded over the last fifty years or so following the cessation of grazing. Examples of such unplanned, spontaneous rewilding include the remote Ewyas Harold Common in the Welsh borders and the tourist hotspot of Dovedale in the Peak District.[56] At a larger scale there is now rapid natural regeneration of woodland on formerly grazed sites in many Mediterranean and East European countries where agriculture is being modernized or abandoned. This unplanned and unprecedented rapid woodland expansion is seen by some as a welcome return to some sort of idealized past before the agricultural and pastoral activities of humans started to manage and mould trees and woods thousands of years ago. It has encouraged the spread of large numbers of wild boar, deer and wolves, which had been erased from the landscape by intense cultivation, trapping and shooting in the nineteenth century. Such unplanned rewilding obliterates traditional agricultural landscapes and may reduce biodiversity associated with the pastures and meadows of the cultural landscape it replaces.[57] But it also allows the reintroduction of rare mammals and birds, the production of some valuable timber and the opportunity for increased carbon sequestration.

3

ANIMALS AND TREES

Woods and forests have long been places dominated by the management of trees and both wild and domesticated animals. For several thousand years wood pastures were vital for local and national economies. These pastures have largely been abandoned and knowledge of how the pasturing of flocks and herds of domestic sheep, goats and cattle within wood pasture can be managed effectively could soon be lost. This is of concern, as such knowledge is required to protect and maximize biodiversity and conserve cultural landscapes. The expansion of forest over the last hundred years or so has provided excellent habitats for several species of wild mammals. There have been rapid increases, for example, in deer populations throughout Europe, which can cause significant damage to young plantations. The wild boar has been reintroduced to many parts of Europe and its recent introduction to England is controversial. This chapter considers the complexity of the relationships between animals, trees and woods, focusing on four animals that have been introduced or reintroduced to the UK and many parts of Europe.

Many woodland landscapes have been created and managed to conserve selected animals, often specifically for hunting. Examples include medieval and more recent deer parks, royal forests and plantations for pheasant shooting and fox hunting. Nowadays considerable efforts are made in the management of woodland to conserve animal species ranging from insects to birds and small and large mammals. Much research has been carried out to understand the life cycles of specific species so that effective conservation management can be undertaken. This includes the creation and management of woodland rides, paths and open areas to encourage butterfly populations. At Haugh Wood in Herefordshire, for example, the woodland tracks are carefully managed to encourage larval foodplants essential for the survival of the rare wood white butterfly (*Leptidea sinapis*). These include leguminous plants such as meadow vetchling (*Lathyrus pratensis*), common bird's-foot-trefoil (*Lotus corniculatus*) and greater bird's-foot-trefoil (*Lotus pedunculatus*). The population of the wood white butterfly is declining in Britain and so considerable efforts are made to ensure that the ride edges do not become overgrown with shrubs and trees. The Haugh Wood management plan aims to increase the quality of ride-side habitat and to widen rides by 'pulling back the conifer plantations and allowing a graduated edge to develop, rich in nectar sources', which will be wide enough not to be shaded by conifers or affected by road management.[1]

One often remembers unusual sightings of animals. An example that springs to mind is seeing a stoat dancing and mesmerizing a rabbit in woodland just to the south of Sheffield about 45 years ago: stoats and weasels are both known for using this technique to enable them to catch their prey.[2] Other examples include seeing long-horned cattle browsing low-growing leaves from ancient oaks in Sherwood Forest and the intense activity of wood ants that will attack a bluebell flower, turning the bluebell pink with their formic acid bites. Perhaps one of the oddest animal sightings in a tree I remember is a toad (*Bufo bufo*) I saw in August 2014. A large branch of an ancient small-leaved lime tree, though still attached, had collapsed and dangled dangerously over a fence. As I cut through the branch with a chainsaw, I noticed to my astonishment a toad actually within the wood of the branch (illus. 17).

Some amphibians do climb trees, of course, including the European tree frog (*Hyla arborea*), which is found throughout much of Europe.

This species favours shrubs and open woodlands with water nearby rather than dense woodland. In Liguria, for example, it is often found in olive groves and thrives in suburban areas such as the coastal resort of Nervi, where the croaking of the males can be almost deafening in the spring. The tree frog may once have been native in Britain and its description by Sir Thomas Browne (1605–1682) in *Pseudodoxia epidemica; or, Enquiries into Very Many Received Tenents and Commonly Presumed Truths* (1646) as 'the little Frog of an excellent Parrat-green, that usually sits on Trees and Bushes, and is therefore called *Ranunculus viridis*, or *arboreous*' is used as evidence to back this up. However, Sir Thomas is most likely to have noticed the tree frog when living on the Continent in Montpellier, Padua and Leiden in the early 1630s. It is possible that there was until recently a native remnant population at Beaulieu in Hampshire. There have been several attempts to introduce it and it is a potential beneficiary of rewilding schemes.[3]

Toads are another matter. It is well known that toads hibernate in damp places and can be found deep in drains and under rocks, but no one I knew had heard of a toad residing in a tree. European toads are not usually associated with trees and there are very few observations of the arboreal habits of toads. A Spanish study found that toads made use of the exposed roots and lower trunks of ancient oak trees in Orgi Wood, Lizaso, Navarra. In 2002–3, 129 observations of toads were recorded with an average climbing height of 39 centimetres and a maximum of 197. They mainly climbed in the dry season and were probably searching for moisture in the moss growing on the trees. They had favourite trees and were 'active all night, and at dawn seek shelter in burrows excavated at the base of the tree'.[4] A fascinating recent study by the zoologist Silviu Petrovan and colleagues found considerable evidence of toads in English trees. Following an enquiry about a toad found in a dormouse nest box in England in 2016, data from dormouse and bat boxes were analysed. They found, to their surprise, that toads 'regularly climb trees in Britain and do so across their active period in the year'. They argue that old trees with hollows near ponds were likely to be particularly attractive to toads, which could climb as high as 2.8 metres up into a tree. The reasons for this are unclear but it could be to find safe places to rest well away from likely predators, especially grass snakes, which prefer open areas to woodland, although they often hibernate in wood banks.[5] So the toad I found within the branch of an old lime tree probably chose its hideaway for its moist coolness and relative safety on a hot summer's day.

There can sometimes be unexpected interconnections between introduced species. One such combination is the attraction that the Asian hornet (*Vespa velutina*) has for eucalyptus trees. The Asian hornet was introduced from China to France in 2004 and has since rapidly spread throughout much of Europe. It attacks bees and other beneficial species and can cause significant losses to colonies of honey bees. The first sighting of the hornet in the UK was in 2016 and by July 2024 there had been 121 confirmed sightings, of which 78 were in 2023. Attempts are currently being made to halt its spread and 86 nests have been destroyed.[6] However, in much of Europe it is now very well established. In areas where there are many eucalyptus plantations, such as Galicia in northern Spain, where Asian hornets were first recorded in 2012, adults are particularly attracted to the sap exuded by recently felled eucalyptus trees and stacks of logs. They kill insects, especially honey bees, and feed them to their larvae for protein.

There is a fascinating link with the European honey buzzard (*Pernis apivorus*). Salvador Rebello and colleagues studied the diet of honey buzzards in Galicia. The honey buzzard is migratory, spending its winters in Africa and breeding in northern Spain in May, staying on until September. This fits in well with the life cycle of the hornets, which become active in May and in June build large secondary nests. Unlike native hornets (*Vespa crabro*) which usually nest within holes in trees, the Asian hornet nests are clearly visible to the buzzards. Trail cameras were used in combination with field surveys of the buzzard nests to calculate the importance of hornet larvae in the buzzard diet. Remarkably, only seven years or so since the arrival of the Asian hornet in Galicia 'it was the second most consumed vespid species' after the common wasp, and 'based on biomass' at some nests it was the most important. The honey buzzard seems to be benefiting from the Asian hornet as the number of breeding pairs increased threefold since its arrival.[7]

Wild boar

The wild boar (*Sus scrofa*) is native to Europe and Asia but now widespread also in North America and Australia, and so successful that in many areas it is classed as invasive. In Australia feral pigs are seen as significant agricultural and environmental pests. At the world scale they have been identified as posing a threat to 672 taxa in 54 different countries, most of which 'are listed as critically endangered or endangered and 14 species have been driven to extinction as a direct result of impacts from wild pigs'.[8] In European history wild boars have long been associated with strength, ferocity and forests. Their destructive habits make them wholly incompatible with settled, domesticated and productive countryside. In the *Iliad* the King of Calydon made the mistake of not offering Artemis, the daughter of Zeus, 'the first fruits of the harvest of his rich orchard plot'. She 'grew angry and sent against him a fierce wild boar, white of tusk', which destroyed the king's orchard: 'many a tall tree did it uproot and cast on the ground, root and apple blossom and all.' Then the enormous boar attacked and killed many men in the region around Calydon in Aetolia, just inland of present-day Missolonghi. The king's son Meleager was the leader of a group of heroes who set out to protect the area in the Calydonian boar hunt. Rubens depicts Meleager placing the final blow on an enormous wild boar, the hunt taking place in a wooded landscape (illus. 37).[9]

In John Gay's fable 'The Wild Boar and the Ram' (1727) a ferocious boar is contrasted with a domesticated ram. The 'savage' boar mocks the cowardly 'fleecy brood' of sheep who 'in silent fright' watch 'the horrid sight' as one of the flock is butchered. The dead sheep is tied up 'against an elm' and 'the butcher's knife in blood was dy'd'.

Samuel Howitt's watercolour *A Boar and a Ram* contrasts the violent and aggressive boar with the placid sheep (illus. 38). The ram replies 'think us not of soul so tame' just because we don't have 'thy tusks to kill'. Eventually, he points out, the sheep get their own back 'Since drums and parchment were invented'. This is because humans used the sheepskin to make the 'two chief plagues that waste mankind': vellum used by lawyers, 'the wrangling bar', and drum skins which wake 'their slumbring sons to war'.[10]

The association between wild boars and royal and aristocratic hunting means that they frequently occur in European literature and art. In practice, their propensity to destroy crops means that over the last 1,000 years their range became circumscribed and restricted to heavily wooded areas in central and eastern Europe. They became extinct in England in the thirteenth century and in northwestern Italy in the early nineteenth century. In recent years, however, rural land abandonment, rural depopulation and the enthusiasm for rewilding has allowed populations of wild boar to spread remarkably rapidly. We studied the causes of this spread and its impact in a remote valley in northwest Italy.[11] Today, when walking in the woods, pastures and small parcels of cultivated land in the Alta Val di Vara, Liguria, one is almost immediately aware of the effects of the feeding habits of the wild boar. Substantial areas of grassland are disturbed by the rooting up of turf by boars in pursuit of worms and grubs; wallowing areas are established in damp areas of woodland and most patches of crops and vines are surrounded by electric fences.

The cultural geographer Robert Hearn interviewed fifty farmers, hunters and local residents in 2010 to unravel the boars' story. Unlike the myth of Artemis and Meleager, the Ligurian boars were not sent by an angry goddess but escaped from a hunting enclosure high up in the Apennines. According to an article published in the *Cronaca della Spezia* on 15 February 1963, 'driven by the heavy snowfall, the wild boar dispersed and escaped from the reserve on Monte Gottero, reappearing in places where their presence is entirely unusual.' The farmers interviewed soon noticed that the boars were damaging their crops and land. A local bar owner recalled that people started to complain about damage in the 1970s or '80s, while a farmer thought the damage had started in the 1960s. The consensus was that the boars had a significant and negative impact on many crops, plants, trees and types of terrain. One farmer said 'It is an invasion! They destroy everything!' Another, that they eat all crops, 'fruit, maize, chestnuts, and what they don't eat, they destroy with rooting for worms and grubs ... I used to grow a number of things, but now I buy them, but I am getting older of course!' For several older farmers the wild boar damage was the final straw that made them decide to stop cultivating the land. In this way the process of unplanned rewilding of the valley, which had started with rural depopulation in the 1950s, and which had provided the habitat to allow the boars to spread so rapidly, was hastened by the activities of the boars themselves.[12]

The boars damaged vineyards and orchards. A vineyard owner found them to be one of the biggest obstacles she faced as a wine producer. The boars were particularly keen on eating the sweet Moscato grapes. In addition they damaged the physical structure of the vineyards: by rooting along the terraced rows of vines, they loosened the soil, damaged the vine roots and weakened the stone and earth terrace walls. She lost between a fifth and a quarter of production due to wild boar, but expensive metal and electric fencing could reduce that loss to a tenth. The boar also harmed the commercially important collection of woodland fungi, especially porcini (*Boletus edulis*). One collector pointed out that the boars tended to focus on areas where porcini were to be found: 'they don't even eat them, they want to find worms, but their rooting disrupts their future growth.' The collapse of traditional chestnut collection was one factor that encouraged the rapid spread of the boar population. A farmer noted that the woods were now 'full of food for them, and so they have more litters, and so cause more damage to cultivation. It is a cycle.' Many people thought that hunting was the only effective control method but some hunters were happy to tolerate the damage caused by the large population of wild boars in order to ensure they had sufficient quarry to hunt. Moreover, as in many parts of Europe, wild boar is an increasingly popular meat served at home and in restaurants.[13]

In the last twenty years wild boars have started to move into European cities such as Genoa, Barcelona, Berlin and Warsaw. For some scientists the boars have become 'native invaders', meaning a species which is becoming invasive within its own natural distribution range. Once the population of boars in surrounding rural areas is high enough, they start to move into the cities. The principal risks include road traffic accidents, potential spread of diseases, occasional physical attacks on people, the spread of rubbish from overturned bins and damage to parkland and gardens.[14] A study in Genoa showed that the boar made use of semi-natural corridors, mainly along river valleys, to move into the city from the surrounding countryside. I have seen boar in the narrow, partially vegetated, riverbed of the River Bisagno, surrounded by city-centre buildings and roads, and boar are frequently observed in the medieval and commercial centre of the city and parks. In recent years increasing numbers of roe deer (*Capreolus capreolus*) have also been seen in the riverbeds and in September 2021 the first wolf (*Canis lupus italicus*) was seen there, most likely attracted by the young wild boar. The study showed that almost 80 per cent of people surveyed had seen a wild boar at least once in their neighbourhood and almost two-thirds of those living in the Val Bisagno had daily encounters with them.[15]

In Britain the wild boar appeared in the wild in the 1980s after a gap of almost seven hundred years. They escaped from several private collections and wild boar farms and semi-naturalized colonies are now found in Gloucestershire, especially the Forest of Dean, Devon, Dorset, Sussex, Kent, Lochaber and Dumfries and Galloway. The large Forest of Dean population, originally made up of escapees from a wild boar farm, became established in the 1990s in woodland near Ross-on-Wye. This population was supplemented by the illegal release 'of around 60 farm-reared animals' near the village of Staunton in 2004 and by '2009 the two populations had merged and a breeding population was thriving'. Their origin as 'farm-bred wild boar' means that Forestry England considers 'they differ from their truly native cousins in other parts of Europe.' The animals are 'less nervous of people, and they are more productive', with an average litter size nearly double that of European boar. Moreover, there are few natural predators in England. This means that the population, which was estimated to be around 150 in 2009, had risen to over 1,600 in 2019.

Once the wild boar successfully became naturalized their legal status was questioned. As former 'farmed animals' they were subject to the Dangerous Wild Animals Act 1976 and owners were responsible for confining them and preventing their escape. Under the Wildlife and Countryside Act 1981 it was an offence 'to release or allow to escape into the wild, any animal that is not ordinarily resident in, and is not a regular visitor to Great Britain in a wild state'. But what was their status once they were freely roaming? DEFRA published a report in 2008 which defined such wild boars as 'feral wild animals, which do not belong to anyone'. The management of these feral wild boars is 'a vexed and contentious issue', as many people delight in seeing the boars, while others wish them to be controlled. The Forestry Commission worked with various groups and planned in 2008 to keep the population at 90 animals by culling. This was raised to four hundred animals in 2013, and the estimated population in 2023/4 was 658.[16] It remains to be seen whether the culling will remain effective within the Forest of Dean, or whether it will halt the rippling out of the wild boar population to areas neighbouring the forest.

Pigs and pannage

Today most people do not associate domesticated pigs with woodland. But in the medieval period pigs were of central importance to European forest economies and various systems of pannage were introduced. Those people with rights of pannage were allowed, often for a fee, to release their pigs in wooded areas to feed on acorns and beechmast in the autumn. The pigs of different owners were usually herded together by a swineherd. Pigs were one of the most important elements of the medieval diet.[17] In the eleventh century pigs were such an integral part of English forests that in the Domesday Book (1086) the extent of woodland in parts of eastern England is listed as 'wood for so many swine' and in southeast England as 'wood rendering x swine annually for pannage'.[18] The number of pigs a woodland could hold, and hence the rent that could be charged, was more important to tax collectors than a wood's area or the value of its timber. Pannage declined in importance from the late medieval period onwards.[19]

In England pannage hung on to a limited extent in the New Forest and William Gilpin (1724–1804) provides a fascinating description of the process in his *Remarks on Forest Scenery* (1791). It was 'among the rights of the forest-borderers to feed their hogs in the forest, during the *pawnage-month*, as it is called, which commences about the end of September, and lasts six weeks'. The swineherd 'collects his colony among the farmers, with whom he commonly agrees for a shilling a head, and will get together perhaps a herd of five or six hundred hogs'. These pigs were then trained to return to temporary, well-watered enclosures in the Forest by being fed with acorns each night while the herder blew a horn. Eventually the pigs got so used to this that after wandering 2–3 miles during the day they always returned at night for their supper: 'Now and then, in calm weather, when mast falls sparingly, he calls them perhaps together by the music of his horn to a gratuitous meal.' Gilpin notes that 'the hog is commonly supposed to be an obstinate, head-strong, unmanageable brute' but that 'if he be properly managed, he is an orderly, docile animal.' Moreover they had 'social feelings' when they are 'at liberty to indulge them' and 'in their daily excursions' they ranged the forest in 'little knots, and societies, as have formerly had habits of intimacy together'. He thought there was something 'uncommonly pleasing in the lives of these emigrants' who could be seen 'going about at their ease, and conversing with each other in short, pithy, interrupted sentences'.[20]

By the late twentieth century Colin Tubbs, a pioneer historical ecologist, lamented the decline in the number of pigs pannaged in the New Forest. In the nineteenth century between 5,000 and 6,000 pigs consumed the acorns in good mast years, but the number collapsed in the twentieth century and by the 1980s was down to a few hundred. In 1986 Tubbs, who had lived in the Forest since the 1950s, noted, 'As I write the acorns are as thick on the ground as I can ever recall them: yet few have thought it worthwhile to turn out pigs to exploit the profligacy.' Pigs are still allowed into the New Forest and in 2023 around two hundred pigs were permitted to eat acorns for a couple of months in the autumn.[21]

How were the pigs managed in woodland areas and combined with other important uses such as hunting and timber? Two documents provide a detailed insight into the management of pigs in Sherwood Forest in the late seventeenth century and the eighteenth. One dated 3 April 1688, an *Agreement between Richard Neale of Mansfield Woodhouse and the inhabitants of Edwinstowe to regulate the number of pigs allowed to eat acorns in Birklands and Bilhaugh*, shows how traditional rights concerning the number of pigs that local people could release into the forest were increasingly codified.[22] These areas of oak and birch woodpasture were part of the Royal Forest and local tenants and cottagers had rights of pannage in them. The document is from the papers of Henry Cavendish, 2nd Duke of Newcastle (1630–1691), who succeeded to the title in 1676. One of the duke's principal estates was Welbeck Abbey, with extensive landholdings in Sherwood. In the autumn of 1688, during the Glorious Revolution, the duke was a key figure in failed attempts to defend the interests of James II against William of Orange and for a time in November 1688 he became a prisoner at his own house at Welbeck.[23] The duke survived the Revolution and remained keen to increase the profitability of his estate and to understand the level of common rights over land bordering his estate in the Royal Forest.

The agreement from his archives follows a dispute with 'Richard Neale of Mansfield Woodhouse in the County of Nott gentleman', who had 'lately purchased to him and his heirs the Herbage and Pannage of the Hays of Burkland and Billhay als Billhaigh in the Forest of Sherwood'. It goes on to state that 'the Men and Remnants [widows] of the Town of Edwinstowe' were 'intending to have right of pannage within the said Hayes for all their Hoggs and Swine without number by virtue of A Grant to them made by late King of England'. This agreement attempts to reduce the pig population: the number of pigs that might roam in the forest for acorns was specifically related to the value of property held by a tenant. Thus 'every Inhabitant . . . who holds . . . a House or Lands above the yearly value of Twenty Shillings shall have the liberty and right only of Pannage and within the said Hays for two Hoggs for each pound rent paying yearly . . . Two pence for the Pannage of every Hogg.' Meanwhile, 'every cottager . . . under the said yearly rent of Twenty shillings have right of Pannage for Two Hoggs only paying yearly for the same two pence the piece.' Finally 'every Cottage uninhabited shall have no right of Pannage for any Hoggs whatsoever.' Those who signed agreed that this 'writing may for ever hereafter Remain as a Rule and stint for Pannage for the Inhabitants of Edwinstowe aforesaid in the said Hays And for preventing all future troubles and controversies about the same'. The signatories included Richard Neale of Mansfield Woodhouse, William Silverton, the vicar of Edwinstowe (1680–99), and thirteen tenants, eleven men and two women; six of the tenants made their mark rather than a signature. Later documents show that keeping pigs at Edwinstowe remained important for the local economy and inhabitants into the early nineteenth century.

The second document is an 'Account of pigs and mast' dated 19 December 1749 in the estate papers of Sir George Savile (1726–1784). Sir George, who had recently inherited the extensive Rufford estate, was MP for Yorkshire, an early supporter of the abolition of slavery and described as a 'great patriot and philanthropist'.[24] He was also keen to maximize his estate income. The document shows that he paid the owner of the neighbouring Welbeck estate, Lady Oxford, for the pannage rights. He in turn sold the rights to those who wished to put their pigs into the forest. The 'mast account' was healthily profitable. The outgoings (rounded to the nearest pound) were £46, the income, 'Money Charg'd for taking pigs into the Mast in Birkland and Bilhay', was £59, making a 'clear profit' of £13 or around 28 per cent. In addition, Sir George gained 105 quarters of acorns worth £21. The payments included £20 for Lady Oxford (of the Welbeck Estate), £5 5s for her agent Mr Wenman, and £5 5s and £10 to the inhabitants of Ollerton and Edwinstowe respectively.

The smaller payments are particularly interesting, as they provide a vivid picture of how pannage worked in practice.[25] A payment was made for 'crying the Mast' at several local villages to advertise that cottagers could pay to put their pigs in the wood pastures. Two men, Abram Jeniver (71 days) and George Johnson (74 days), were paid 1s a day to work as 'tenters', which involved looking after the mixed herds of pigs. The pigs were branded to identify their owners and Abram was paid 1s 8d to do this; Abram also received payment for 'Ale at several times' amounting to 6s 10d. Sometimes pigs escaped from the tenters and 5s were paid for 'swine being pounded'. The owners of pigs were paid if their pigs died while being looked after by the tenters, although this was rare: in 1749 William Thompson was paid 14s for '2 piggs that Dyed', and Mr Clark was paid 3s for 'One that did not thrive'. This profitable form of pannage died out in the early nineteenth century and by the 1860s turning swine into the forest was only a memory for the oldest inhabitants. Today the area known as Birklands is the core of the Sherwood Forest National Nature Reserve. Knowledge of past land-use management practices such as pannage is essential for developing plans for reintroducing domestic stock to help conserve the characteristic heathland and birch and oak woodland and the species they sustain.

The Hungarian ecologist Zsolt Molnár and colleagues have worked closely with traditional Serbian swineherds (svinjars) in Bosut Forest, which lies on the borders of Serbia, Croatia and Bosnia. It is an oak, ash, hornbeam and maple floodplain forest alongside the River Sava. Before the mid-nineteenth century the old-growth oak forest produced timber combined with grazing and pannage. Pigs were usually kept out of compartments where oak was being regenerated, and allowed back in when 'the stand starts producing acorns regularly', usually after fifty years of growth. The number of pigs and svinjars has been in decline for years and the research captures traditional knowledge which is likely to be lost. The svinjars are 'observant people' whose 'main source of knowledge is direct observation of pigs and nature' over many decades. They 'speak from experience, having lived through forage-rich and forage-poor years and seasons'. They have observed relatively rare events such as floods and particularly severe winters, and, longer-term, more gradual environmental changes. The svinjars follow many traditional practices, but old pig breeds such as Sremska Lasa and Manulica have largely been replaced by hybrids of Yorkshire, Duroc, Landrace and Piétrain breeds. Most keep between 50 to 150 pigs covering an area of 30–100 hectares (illus. 19). Svinjars pay the forestry authority a fee, which is higher when it is a good mast year. Since the reappearance and rapid spread of the golden jackal (*Canis aureus*) in the twentieth century 'most svinjars only keep their 10–30 sows in the forest, while piglets are brought to the village after weaning.' Maize is the principal fodder provided by the svinjars for fattening their pigs and helping them to know where home is.[26]

If pigs are kept out of young stands of oak effectively there appears to be a healthy relationship between the pigs and oak regeneration. Svinjars consider that pigs loosen the soil and reduce the growth of competing vegetation: one retired forester noted that 'a pig eats up one acorn while treading two into the soil.' The svinjars had a detailed knowledge of which plants pigs preferred to eat, and identified 98 species that were eaten, and 56 which were ignored. The acorns of *Quercus robur* were the pigs' favourite food. They 'search for early fruiting trees (in early September) and eat acorns in the morning and evening for about half an hour', while ripe acorns are 'eaten all day'. As the winter progresses the acorns become sweeter and in January they may be eaten from below the snow. They forage in small groups of up to three pigs and 'know where the sweetest acorns are (always from the same trees)'. Pigs can eat up to 5 kilograms of acorns a day, washed down with water or snow.

Until the 1950s acorns were collected by svinjars and stored in pits; the germinating acorns were then fed to the pigs from December onwards. The researchers found that 'pigs alternated between eating acorns and earthworms several times during the day'. Indeed, earthworms were their favourite animal food, and it is estimated that a single pig eats between 400 and 1,000 worms 'in an active day in February and March . . . Pigs slurped them up (like children eating spaghetti)'. This was especially true in years when there was a shortage of acorns. Another favourite forest food is porcini, which they prefer over acorns and even over maize: 'they will not come home, they sleep in the forest in porcini season.' During the spring the pigs browse foliage, favouring young hawthorn, hornbeam and maple leaves. This Serbian research confirms that pigs are omnivorous yet 'highly choosy, and could even be described as gourmets'. If carefully managed, pannage can be combined with other forest uses, and allow regeneration of young trees. It can benefit forest diversity and has a positive impact on the health of the animals and the meat produced.[27]

Beavers

The European beaver (*Castor fiber*) was once found throughout Europe and across Asia to Siberia. Its population was much reduced by hunting for its fur during the medieval period and improved traps and guns caused a rapid decline from the seventeenth century, leading to its extinction in many European countries. By the start of the twentieth century it is estimated that only 1,200 beavers survived in eight enclaves in France, Germany, Norway, Ukraine and Russia. The species was legally protected in many countries such as Norway in 1845, France in 1909 and Germany in 1910, but this protection did not appear to have much effect. The first of various national reintroduction campaigns took place in 1922. The idea was to boost beaver populations to provide more fur, and in some cases North American beavers, which were mistakenly thought to be the same species, were introduced.[28] More recently, and especially since the 1970s, there has been pressure to reintroduce beavers to increase the number of large mammals. Beavers are increasingly celebrated as stream and river engineers who, through the building of dams, lodges and the excavation of channels, can have a positive impact on the biodiversity of riparian meadows and woodlands.[29]

Poland has seen a resurgence in the beaver population, which had collapsed during the nineteenth century. After the Second World War some beavers were introduced from Voronezh, USSR, to try and establish a population, but this was largely unsuccessful until the launch of the 'Active Protection of European Beaver in Poland' programme in 1974, which involved both animal scientists and hunters. The impact was remarkable and the beaver population rose from 1,000 in 1977 to around 125,000 in 2017. What was the effect of this dramatic increase on trees? A major problem was 'the destruction of old trees which are particularly valuable in nature and are impossible or very difficult to replace'. In addition trees were sometimes felled across roads or railway lines. A study carried out in 2014 showed that the most common form of reported damage was flooding (46 per cent), followed by trees gnawed and cut down (27 per cent). Although in Poland the beaver is partially protected, some hunting by landowners is allowed if there is significant damage and landowners can also be paid compensation.[30] The historian Melchior Jakubowski took me in September 2022 to see the extent of beaver damage to trees near the village of Sikory on the River Narew, northwest of Warsaw (illus. 18). The floodplain of the Narew consisted of many abandoned meadows near which grew some large aspen, elm, oak and white poplar (*Populus alba*). The photograph shows an elm tree that had been felled by beavers the year before and is resprouting. Nearby there were several heavily gnawed white poplars and other trees which had been ringbarked and remained standing, dead. Several trees had been individually protected by wire fencing to stop the beavers felling them.

Scotland is at a much earlier stage in the beaver reintroduction process than Poland. The last record of beavers was in 1526 when they were noted at Loch Ness. It was to be almost five hundred years before the Scottish government gave formal approval to allow beavers to remain in two areas, Argyll and Tayside, where there were small existing populations. This approval, in November 2016, 'was a historic moment: the first formally approved reintroduction of a mammal species anywhere in the United Kingdom'. Knapdale in Argyll, western Scotland, was the site of the carefully planned Scottish Beaver Trial, which studied sixteen beavers in five family groups released from 2009 to 2011 and their impacts on woodland, water habitats and other species. The Tayside population, in contrast, originated in escapes or deliberate releases from private collections. From 2012 this population was monitored by the Tayside Beaver Group.[31] The trees and woodland most likely to be affected by the spread of beavers are those growing alongside rivers and streams. A study of riparian willow and aspen woodland showed that there was rapid regrowth of these trees when felled by beavers. The willows regrew from cut stumps and the aspen suckered vigorously: 'the conservation status of both these trees or the wider habitats that they form would not be threatened by a reintroduction of beavers to Scotland.'[32]

In England several small populations of beavers are now established and there is great interest in the potential for beavers to assist with rewilding projects such as that at Ennerdale in the Lake District, where it is proposed 'to see free living beavers reintroduced to the length of the River Ehen, from mountains to sea'.[33] In 2020 the government announced that the small beaver population on the River Otter in southeast Devon 'could legally remain'. The opportunity was taken to carry out research on the impacts of beavers. This found that some trees were damaged, including the felling of a Bramley apple tree, several poplar trees and a 'willow tree of sentimental value'. Those people affected were helped by Devon Wildlife Trust to fence individual trees and groups of trees to prevent future damage.[34] It is likely that beaver management groups of local residents, farmers, woodland managers and conservationists will be a vital approach to mollify the impacts of beavers as they spread more widely.[35]

Reeves' muntjac deer

The history of European forests is strongly linked to that of deer: deer hunting was one of the principal reasons for the protection of forests in the medieval period. Recent archaeological studies on the European fallow deer (*Dama dama*) show the complexity of their relationship with humans. Research 'combining zooarchaeological and biomolecular analysis of ancient and modern *Dama* remains' has allowed the movement of the species from its glacial refuge in Anatolia to be traced. Studies of their bones show that they were frequently associated with domesticated animals such as cattle, goats, sheep and pigs, whose spread across Europe started in the Neolithic period. The deer was strongly linked with Artemis, the Greek goddess of hunting, and Diana her Roman successor, and the popularity of Artemis with Greek settlers in Sicily is indicated by archaeological finds. During the Roman period fallow dear spread rapidly through the empire and were linked with Diana and hunting in parks. At Fishbourne Roman Palace, Sussex, fallow deer remains have been dated to the first century AD, soon after the invasion, and there is also evidence of their breeding in later years. Fallow deer died out in Britain after the end of the Roman period, and were reintroduced – not, as has long been thought, by the Normans from Sicily, but around AD 1000 from Anatolia. Fallow deer parks were an important statement of power for elite landowners in Norman England, and English deer were exported to Ireland, France and several northern European countries.[36]

In broad terms, agricultural modernization and the decline in the social and cultural importance of deer hunting led to declines in deer populations until the mid-twentieth century. Subsequent increases in woods and forests have allowed very large increases in native and introduced deer populations. In Britain by the end of the nineteenth century there were only small remnant populations of two native species: the roe (*Capreolus capreolus*) and the red (*Cervus elaphus*). Today the populations and ranges of these native deer have spread, and they have been joined by four introduced deer: the fallow (*Dama dama*), Sika (*Cervus nippon*), Reeves' muntjac (*Muntiacus reevesi*) and Chinese water deer (*Hydropotes inermis*). The different species affect woodland regeneration and ecology in different ways. The roe expanded from a very low population in the early twentieth century and is now 'the most widely distributed deer in Britain'. They browse leaves and bramble is a favourite: you often know that roe are present in a wood by a combination of the hoof prints, or slots, they leave in muddy rides and fact that the tips of many bramble shoots have been eaten.[37] Deer densities 'are at extraordinarily high levels' and caused by several factors including 'the absence of large predators, a decline in hunting pressure, concentrations of populations in habitat fragments and the widespread invasion of non-native, smaller deer species'.[38]

Reeves' muntjac, a small and initially unobtrusive deer, quietly spread through the countryside during the twentieth century until it became common in many parts of England and a significant threat to woodlands. There is confusion over the naming of this slightly humpbacked deer, which is native to parts of southeast China and Taiwan. Some say the deer was collected by the amateur naturalist John Reeves (1774–1865), who became the chief tea inspector for the British East India Company at their Canton factory. He arrived in China in 1812 and had been advised by Sir Joseph Banks on the best ways to collect natural history specimens. Reeves also commissioned local artists to draw birds, fish and plants and sent many plant and animal species, live and preserved, to experts in Britain, while based in China. These included the exotic Reeves' pheasant (*Syrmaticus reevesii*), which became a popular aviary bird in Victorian Britain although attempts to natural-ize it were largely unsuccessful.[39] However Reeves retired to live in Clapham in 1831 and it was his son John Russell Reeves (1804–1877), also based in China and a keen natural history collector, who pre-sented a pair of muntjac to the Zoological Society of London in 1836. This pair was supplemented by other deer and the species bred successfully fourteen times between 1874 and 1881.

The history of the muntjac's introduction is complicated, with moves from zoological collections to private collections in parks often not fully documented. Some of the muntjac from London Zoo were sold to the Jardin des Plantes in Paris and a London animal dealer, William Jamrach, purchased several muntjac between 1883 and 1902 from the group at Paris. Jamrach was the son of Charles Jamrach, who from the 1840s had established a leading import and export trade in exotic animals in premises known as Jamrach's Animal Emporium on Ratcliff Highway in the East End. In 1902 William Jamrach sold a male and two females to Herbrand Arthur Russell, 11th Duke of Bedford (1858–1940), at Woburn: 'it seems probable that at least some of the animals supplied to Woburn were the descendants of stock from London Zoo.'[40] The duke's interest in muntjacs was part of a much wider interest in the conservation of mammals: he was an enthusi-astic naturalist, conservationist and zoologist. He was president of the Zoological Society of London from 1899 to 1936 and the links between his private estate at Woburn and London Zoo were strong. He introduced Przewalski's horses and the extremely rare Père David's deer to the zoo. After the First World War he supported the establish-ment of Whipsnade Park Zoo, opened in 1931, and to commemorate this he presented a pair of muntjac together with several other deer for the park.[41]

Another extremely wealthy landowner with an interest in munt-jac was Lionel Walter Rothschild (1868–1937), who became 2nd Baron Rothschild in 1915. He was an ardent zoologist who travelled in Europe and North Africa, employed explorers to collect zoological specimens and established a museum at Tring Park, Hertfordshire, now part of the Natural History Museum. A female muntjac was given to the enthusiastic young naturalist by an uncle in January 1881, probably as a present for his bar mitzvah. There are several early records of muntjac escapes from Tring Park in the 1930s but it is not known whether these helped to form part of a 'free-living population'. Detailed records concerning the muntjac at Tring were destroyed in the Second World War.

At Woburn, luckily, records survive for muntjac kept there in enclosures from 1894 to 1903, although these were fragmentary until Mary, Duchess of Bedford, a keen ornithologist, began compiling exact records in 1896. These show a small breeding population of between seven and thirteen deer, supplemented by some purchases: the population survived but did not thrive. In March 1901 eight males, two females and one juvenile were 'released into the woods' and there was a 'no culling policy'. The population gained ground and extended its range outside the park, although they were not 'particularly common in Bedfordshire even by the early 1970s'. From this core area muntjac began to spread quickly and by the 1990s were found through the southeast and the Midlands. Although their rapid spread was of considerable interest to zoologists, as late as 1994 their impact on trees and woodlands was still uncertain. The key review paper noted: 'Although Reeves' Muntjac apparently could have a substantial ecological impact, there are few data on their habitat preferences, their likely rate of further spread, or future population increases.'[42] Yet within a few years the adverse impact of muntjac on the conservation value of many woods became obvious and an important 2001 study at Monks Wood National Nature Reserve concluded that many 'conservation woodlands require protection of coppice to try to prevent unacceptable levels of damage'.[43]

Muntjac are now very common in much of England and in some well-wooded parts of southeast England there may be as many as 110 individuals per square kilometre, although as they are hard to spot it is difficult to make accurate estimates. They are secretive and are most commonly seen in the light from car headlamps. They form tunnels through undergrowth of brambles and can move around recently felled woodland without being seen. When disturbed they bark, making a sound like a yapping dog. Muntjac are usually seen by themselves or in small family groups; they do not form herds. Bucks are territorial and mark tree trunks and boughs with scent from glands on their forehead. Muntjac breed all year round. The female produces around three fawns every two years and adults live for between ten and thirteen years.[44] The increase in muntjac is making the management of woodland and regeneration of trees increasingly difficult. The regrowth from coppice is constantly nibbled and eaten and hence is halted and much reduced. Young saplings are eaten back. The Mammal Society notes that 'Because they breed all year, it is difficult to suggest a sensible humane culling season; and because they are so small, and often in suburban habitats, shooting is difficult. Bucks can be shot at any time, as can young does, before they have reached sexual maturity.'[45]

The implications for nature conservation and woodland management are serious. One of the most successful conservation policies in recent decades has been the reintroduction of coppicing to encourage displays of spring flowers such as bluebells, wood anemones and primroses. This management has also allowed the production of valuable firewood from the coppice regrowth. Now that muntjac are so common, such coppicing is likely to be increasingly uneconomic as the regrowth will be much reduced. Muntjac are brawny and can easily push under, and leap over, light fences. Strong deer fencing may be the only option, although this will make small-scale coppicing uneconomic. Another approach is to protect areas of coppice regrowth with dead hedges made of brash. In addition, muntjac enjoy eating wildflowers such as primroses and orchids. Unlike other deer, they also eat bluebells, which have been one of the main beneficiaries of the reintroduction of coppicing. I have noticed that they also eat herb

12 Benjamin Pouncy, after Thomas Hearne,
An Oak in Moccas Park, Herefordshire, 1798, etching.

13 Unknown photographer, *Two European Men Standing by a Canoe
Tree in the Murray River Region*, 1914, photograph.

14 Velebit mountains, northern Dalmatia, July 2016.

15 Old pollarded oak, Vrsno, Šibenik, July 2019.

16 Sitka spruce plantation, Perthshire.

17 Toad within lime tree branch, Aylton, Herefordshire, August 2014.

18 Beaver damage near Sikory on the River Narew, northwest of Warsaw, September 2022.

19 Pigs rooting in Bosut Forest, Serbia, 2019.

20 Dead elm hedgerow, Westhorpe, Southwell, 1984.

21 Active management of chestnut as coppice, Costola, Liguria, April 2015.

22 Effects of fire on karst near Vrsno, Šibenik, July 2019.

23 Palms and gardens, Val di Sasso, Bordighera, September 2018.

24 Rotunda at Croome Park, Worcestershire,
encircled by cedars of Lebanon, July 2022.

25 Glastonbury thorn, Cradley Churchyard, Herefordshire, 2023.

26 *Pinus pinea* growing among the imperial ruins of Rome.

27 Rewilding at Ennerdale, Forestry England, May 2009.

Paris (*Paris quadrifolia*), which can be a valuable indicator of ancient woodland. One of the few advantages of muntjac is that they often make paths through thick regrowth of bramble, which makes it easier for humans to walk through the dense undergrowth. In addition, they are short, and so cannot reach up to browse above a metre or so. However, they are intelligent and have learnt to 'walk over' young trees, pushing down the tops with their breasts, which allows them to eat with relish the growing stems of the young trees.

The muntjac is beginning to spread in many European countries. Those released in France in the late nineteenth century failed to naturalize, but they are now reported from several parts of the Loire valley. In Ireland they were first reported in 2007 and they appear to have been released in several sites in Wicklow and Kildare. There has been an increase of sightings in Belgium since 2005, especially in East Flanders and around Antwerp. The first sightings of muntjac in Germany were made in 2004. It is likely that the spread of the muntjac is encouraged by hunters. In the Netherlands muntjac have been reported since 1997, but sightings are not common and since 2000 'the trade and possession of any species has been forbidden.' A recent review of the distribution of the muntjac concluded that 'it could expand its range to include every European country' and that to prevent its adverse effect on conservation interests 'national administrations should consider eradicating Reeves' muntjac while it is still feasible.'[46] It remains to be seen whether this hard-line approach is followed.

4

TREES UNDER ATTACK

S ome of the greatest threats to woods and forests come from tree diseases, pests and fire. Devastating fires occur every year in Mediterranean forests, especially in Portugal, southern France, Greece, Italy and coastal Croatia. Some of these occur naturally following lightning strikes; others are caused by arsonists and human error. There is no doubt that human-induced climate change is worsening the situation and causing the intensification of wildfires, which will become more common, increasingly severe and tend to cover larger areas.[1] This chapter explores the ways humans have helped to increase threats to trees and forests.

There has been a remarkable growth in the number of tree diseases and pests over the last fifty years or so. The Forestry Commission in 2024 lists alphabetically 25 serious tree diseases ranging from acute oak decline to *Xylella fastidiosa*, and 24 tree-attacking insects from the Asian longhorn beetle to the two-spotted oak buprestid beetle (*Agrilus bigattatus*).[2] Some of these diseases have significant impacts on commercial forestry. Red band needle blight, for example, caused by the fungus *Dothistroma septosporum*, attacks pine trees including Corsican pine, Scots pine and lodgepole pine, which are the principal commercial pine trees grown in Britain. The fungus causes pine needles to turn red and die, which weakens trees over several years and eventually kills them. The disease only became significant in Britain in the late

1990s; before that it mainly affected Monterey pines (*Pinus radiata*) growing in the southern hemisphere. By 2006 almost three-quarters of all stands of Corsican pine in England, Wales and Scotland were infected by the disease. Corsican pine was one of the most productive commercial conifers in the UK, thought to be 'well adapted to the climate changes predicted for southern and eastern Britain over the 21st century', but because of the disease is no longer planted.[3]

Phytophthora ('plant destroyer') has been called 'arguably the world's most historic and economically significant genus in plant pathology'. It was first recognized in 1876 when *Phytophthora infestans* was identified as the cause of the potato blight that caused devastation in Europe and especially Ireland in 1845. The number of described species of *Phytophthora* has increased rapidly since the turn of the twenty-first century because of improved techniques and fieldwork in remote forests, and it is now estimated that there are between two hundred and six hundred species.[4] One of the most potent species is *Phytophthora ramorum*, which infects several commercially important trees. It was first identified in Europe in 1993 and in the UK in 2002 on a viburnum plant in a Sussex garden centre and in 2003 on a mature southern red oak (*Quercus falcata*). Six years later it was found on a Japanese larch in southwest England. Since 2009 large numbers of the three commercial species of larch have become infected. It has spread throughout much of western England, Wales and Scotland. The rapid spread appears to have been helped by the fact that larch plantations are frequently associated with the invasive shrub *Rhododendron ponticum*, which is itself highly susceptible to the disease. There is no cure and the current policy is to fell all infected larch trees and nearby trees even though they show no symptoms. Worryingly, the disease was found to have infected several sweet chestnut trees in southwest England in 2015 and further research is being undertaken to discover how this has happened.[5]

Other species of *Phytophthora* are also active. *Phytophthora alni*, first discovered in 1993 in the UK, is now widespread and is estimated to affect around a third of all alder trees. I remember noticing dieback of alders in the 1990s, and the killing of several alders, including mature trees over a hundred years old. Many alders seem to survive even though they grow next to infected trees showing dieback. Infected trees that are coppiced often regrow successfully. It may be that the spread of the disease has been under-reported because alders are more frequently found along river banks and in small farm woodlands than in commercial plantations. In Europe there are high levels of infection in Bavaria and northeastern France.[6] A very recent species is *Phytophthora pluvialis*, which was noticed affecting western red cedar (*Tsuga heterophylla*) in southwest England in August 2021. This species was first identified in southern Oregon in 2013 and then in New Zealand in *Pinus radiata*. One theory is that it was transported to New Zealand in forest machinery before 2001.[7]

There are also significant threats from insects. The emerald ash borer (*Agrilus planipennis*), which is native to east Asia, was identified in 2002 as causing the death and decline of many ash trees around Detroit, and is now common across large parts of the USA and Canada. It has also spread west in Asia, reaching as far as Ukraine in 2020. It can spread distances of 40 kilometres per year and there are concerns that it could spread rapidly west through Europe.[8] One of the most noticeable invasive pests is the horse chestnut leaf miner moth (*Cameraria ohridella*). The larvae of this moth tunnel through the leaves of the horse chestnut (*Aesculus hippocastanum*), disfiguring them and causing large brown patches. Adult moths appear in April and there can be five generations per year in the UK. The moth was first noticed in northern Greece in 1984 and identified as a new species in 1986, although it is possible that leaves with similar mined patches survive in herbaria from the late nineteenth century. The insect rapidly spread through western Europe, reaching Wimbledon in 2002. As the tree is commonly planted in gardens, parks and streets for its flowers and conkers, the insect caused a great deal of concern wherever it appeared. However, it was soon realized that although attacks weaken the trees, most survive the condition. Moreover, its spread could be reduced if all fallen leaves were removed and destroyed. There are still worries that in the long term the moth attacks will weaken trees and encourage trees to shed branches.

The spread of many tree diseases and pests has been brought about by the massive increase in global trade over the last 150 years in wood and timber products. More recently there was a substantial growth in international trade in young trees and saplings. These trades have undoubtedly caused the spread of many diseases. The first part of this chapter considers the complicated interactions between humans, trees and three of the most virulent tree diseases: Dutch elm disease, ash dieback and chestnut blight. It concludes with an assessment of the historical and cultural context for the recent increase in fire risk mainly in Mediterranean Europe.

Elm diseases

Those who experienced the devasting effect of Dutch elm disease in Britain in the 1970s find it hard to forget. However, when the first accounts of the arrival of this disease were reported, it was difficult to imagine that the scale of the impact would be so great. I first saw the disease in Herefordshire, indicated by the yellowing of the leaves of hedgerow elms, in 1971–2 and within a couple of years all the hedgerow elms (*Ulmus procera*) and most of the elms growing in woodland, mainly *Ulmus glabra*, in the district were dead. If there was a slight breeze when walking through woods, there was a characteristic and strangely threatening clickety-clacking of the dead branches against each other. The diarist James Lees-Milne noted on 24 August 1973 at Ragley Hall, Warwickshire: 'The elm disease hereabouts is terrible. Every elm tree large and small, old and young, is a dead skeleton . . . I don't think enough fuss is made about this disease, which is changing the entire character of the Midlands.'[9]

A key reason for not 'enough fuss' being made is that earlier forms of elm disease had long been known in Britain. The elm bark beetle (*Scolytus destructor*) was described in the standard text *Forest Entomology* of 1908 as 'a dreaded pest in elm-growing districts, as for example in many parts of the south of England and London parks'. The leading economic entomologist of the late nineteenth century was Eleanor Ormerod (1828–1901), who virtually single-handedly invented the subject in Britain and was later celebrated by Virginia Woolf in the short story 'Miss Ormerod' (1924) (illus. 28). In her *Manual of Injurious Insects* (1890) she emphasized the destructive impact of the elm beetle, arguing that when an attack was spotted it was vital to remove 'all centres of infestation, from which the beetles might spread to the sound trees'. She was horrified by the 'inexcusable practice . . . of leaving trunks of infected elms, lying, with their bark still on', which contained 'myriads of these maggots, which are all getting ready shortly to change to perfect beetles, and to fly to the nearest growing elms'. She had seen 'neglected trunks' in 'our parks and rural woodyards all over the country'; and had herself been able to

28 Miss Eleanor A. Ormerod.

run her hand 'under the bark, so as to detach feet and yards in length from the trunk, all swarming' with maggots.[10]

The first sightings of another form of elm disease in Europe were in Picardy in 1918. At first it was thought the yellowing of the leaves could be caused by toxic gas or some effect of the First World War, but the symptoms were soon recognized as those of a new elm disease and Dutch scientists successfully identified the pathogenic fungus. The disease has subsequently always been called Dutch elm disease. A Dutch tree recorded with the disease in 1919 was felled and tree ring evidence suggested that the disease was present in 1917; it is likely that it was introduced from Japan to Europe some years earlier.[11] The disease was first recorded in Britain at Totteridge, Hertfordshire, in 1927. In 1926 a row of around fifteen roadside elms had died, and been felled and removed. In 1927 a tree at Totteridge golf course also had yellowing foliage and fungus was successfully cultured from it, so this was the first proven case. Within a couple of years the link between the fungus *Graphium (Ceratostomella) ulmi* and the elm beetle which carried it from tree to tree was established.[12]

The implications of Dutch elm disease were soon recognized: *The Times*, under the headline 'A SERIOUS MENACE', noted on 26 May 1928 that 'When travelling in Holland of late years no one interested in trees can have failed to observe the pitiable condition of so many of the elm trees.' The disease attacked all trees 'irrespective of age and vigour' and 'in Holland no species of elm seems immune.' Now that the disease had gained a foothold at Totteridge, and 'in view of the rapidity with which the epidemic has spread over a large part of the Continent', the article argued that 'the situation is pregnant with disturbing possibilities.' If the spread of the disease was not halted by the felling of the diseased trees then 'Complete destruction of affected trees may prove the wisest policy.' Letters to *The Times* the same week showed that the disease was spreading fast. On 29 May 1928 Robert Adkin wrote, 'Here in Eastbourne, where many miles of the roads are planted with Cornish Elm (*Ulmus stricta*), the disease has appeared in two separate places in the town.' He felt that it was 'desirable to destroy utterly any trees that show the slightest traces of the disease'.

On 30 May 1928 *The Times* reported 'ACTION BY FORESTRY COMMISSION', which 'will next week begin a survey in Hertfordshire of elms likely to be affected' to obtain data on the disease which 'was still somewhat of a mystery to mycologists'. This would be a relief to 'Nature lovers in general, and private owners of elms in particular'. It was felt that an 'Order of the Ministry of Agriculture issued in December 1926, prohibiting the importation of any living elm trees into this country had done much to keep out the disease'. However, Sir George Courthope, representing the Forestry Commission, told the House of Commons on 19 July 1928 that 'Dutch elm disease had been discovered in Hertfordshire, Middlesex and Norfolk. Suspicious cases were also being investigated in Essex, Sussex, and Worcestershire. The research officers of the Forestry Commission had the matter in hand.' *The Times* reported that 'Continental mycologists have now arrived at a definite conclusion as to the cause ... A fungus – *Graphium ulmi* ... when an elm branch is intentionally infected with the fungus, death of the branch and infection of other parts of the tree soon follow.' In an editorial it stressed that 'Anything which affects the elm trees of this country is of national concern' and, moreover, the 'alarm will be all the greater because, as with foot-and-mouth disease in cattle, science can suggest neither prophylactic nor cure'. It was unsure how the disease had arrived but 'interest ordinarily attaching to such questions is sunk in the greater emergency. The disease is here and must be faced.'[13]

The Forestry Commission research undertaken by Tom Peace of the Imperial Forestry Institute, Oxford, indicated that Dutch elm disease was present in the Isle of Wight and other districts before 1927, and by the end of 1929 'the disease was known to be present in 19 counties, restricted to the Midlands and South of England.' It also found that 'The intensity of the disease varies greatly in different districts; as a whole the proportion of trees affected is quite small, and of these not many (usually less than 5 per cent.) have been completely killed.' The most important carrier was identified as *Scolytus destructor,* 'the tunnels of which are nearly always present in the trunks of trees attacked by *Graphium ulmi*'. The Forestry Commission continued its survey every year and by 1932 reported a 'reduced virulence of the attack in nearly every area visited'. However, 'in many of the "recovered" trees the fungus is still alive and experience has proved that a recurrence of the attack is possible.' It was no longer felt necessary to cut down affected trees 'unless they become unsightly or show no signs of recovery': 'The malady is too extensive in England to allow of its eradication being considered a practicable measure.'[14] The spread of the disease fluctuated in the 1930s. In 1935 it was noted that 'after three years in which the disease has been more or less static its incidence is again approaching that of the peak year, 1931.' In Worcestershire '500 trees have been doomed to the axe on one estate.'[15] By the late 1930s the very success of the disease meant that there were fewer trees available to be attacked. In 1938 'A slowing up of the disease . . . has already been noticed in Essex and Hertfordshire where it was formerly very severe, killing more than half of the elms in certain parts of these counties.'[16]

Concern about Dutch elm disease waxed and waned over the next thirty years. The Oxford botanist Nicolas Polunin pointed out in 1943 that 'In many parts of the country, particularly towards the south east, large numbers of elms of both our major species are dying of Dutch elm disease,' and that owners 'should cut down forthwith all dead and obviously dying trees, thereby also creating a considerable store of valuable firewood' useful for the war effort.[17] After the war *The Times* emphasized the effect of the disease on public parks. On 29 December 1953 it showed a photograph of 'The Broad Walk, Kensington Gardens, looking north, as it appeared yesterday, showing elm trees flanking the walk which have been felled in the interests of public safety. So far no tree has been found free from Dutch elm disease and over half were in a dangerous state.' Another picture showed the same part of the Broad Walk 'photographed early in the month, showing the trees standing'. But a certain complacency developed about the fate of hedgerow and woodland elms. It argued that 'The ravages of elm disease – now so distressingly apparent in Kensington Gardens – are unlikely to dislodge the elm from its favoured position as a hedgerow tree.' This was because 'young elms are always arising from suckers and in many areas have partly replaced the losses from disease.' Moreover, the old elms 'heavily lopped through the years, can disappear from a farmer's hedgerow and its place be fairly quickly filled by new growths without much notice being taken, even locally.'[18]

A comprehensive review of Dutch elm disease published by the Forestry Commission in 1960 reported that overall 10–20 per cent of elms had been killed by disease, but that 'this is being made good by the growth of young sucker elms in hedges.' It thought that it was safer to plant elms in cities than in the countryside as it was possible that the elm 'beetles will not tolerate polluted conditions'. It argued that preventative felling of diseased trees was not appropriate because 'It is quite certain that had the felling of diseased elms been practised on a large scale in Britain our losses, resulting in this case indirectly from the disease, would have been much higher than the direct loss that has occurred.' The report concluded that

> The disease may long continue as a minor nuisance,
> but unless it completely changes its present trend of
> behaviour it will never bring about the disaster once
> considered imminent. Indeed in order to wipe out most
> of the elms in Britain it would have to achieve an even
> higher level of virulence than it did in its worst years
> in the nineteen-thirties.[19]

Unfortunately, this is exactly what happened eight years later. The Forestry Commission reported that, starting in 1968, 'a number of serious outbreaks developed causing devastation on a scale greater than ever before and in consequence the disease is once again a cause of great concern.'[20] When members of the public started to report the spread of the disease in the late 1960s officials frequently thought that this was 'merely a flare-up of an already established, but less virulent, form of the disease'. But this was far from the truth and the 1970s outbreak 'stands out as one of the most dramatic environmental events of the last forty years, killing almost 30 million trees and restructuring the traditional "elmscapes" of large parts of lowland England, Scotland and Wales'. The new very virulent strain was introduced from North America to the UK on a consignment of rock elm (*Ulmus thomassii*). In October 1971 steps were taken to enforce felling to prevent the spread of this devastating disease and local authorities were given statutory powers 'to enter land and inspect elm trees' and, if necessary, destroy 'trees without compensation'. However such felling generally only took place where public safety was a factor, and 'no sanitation felling' took place over 'large swathes of rural England'. By the mid-1970s it became obvious that the disease was unstoppable, and the Forestry Commission advised the government that 'there is no alternative to allowing the disease to take its course.' It was recommended that there should be 'a sustained programme of planting hardwood trees other than elms over the next few years'.[21]

The destruction of elm trees across Britain during the 1970s continued throughout the following decades, and the disease remains very active today. The English elm (*Ulmus procera*) still regrows by suckering, but this regrowth is attacked as soon as it reaches an age, around fifteen to twenty years, when it is attractive to elm bark beetles (illus. 20). Within woodland older wych elms (*Ulmus glabra*) can be found. The wych elm spreads by seeding and natural regenerated seedlings are frequently found, but this species is also susceptible to attack. The loss of so many large hedgerow trees encouraged the removal of many hedges. While heavy tree-removing equipment was on farms it was a relatively easy matter to arrange for contractors to remove unwanted hedgerows.

There were few positive outcomes of the disease. It did encourage investment and research in the study of forest pathology by the

29 David Gentleman, special issue of stamp showing oak
Quercus robur, 1973, First Day Cover designed by Peter Hatch.

Forestry Commission. The omnipresent rows of dead and dying trees
also stimulated public interest in the planting of broadleaved trees in
the devastated countryside and the creation of 'new agricultural landscapes'. It also encouraged many tree-planting schemes in villages,
towns and cities. This was reinforced by the 'Plant a Tree in '73' campaign, which was instigated by the government and supported by a
wide variety of organizations. The Post Office issued commemorative
stamps (illus. 29) and the campaign led to the establishment of the
Tree Council, a charity devoted to the planting and conservation of
trees. Perhaps most significantly, the disease focused the attention of
foresters and policy makers on the need for research on broadleaved
woodland management: the influential Broadleaves in Britain conference of 1981 was held partly because of 'a sharpened appreciation'
of the values of the broadleaved tree in terms of 'visual beauty and its
unique variety of wildlife' and value for timber production, recreation
and cultural heritage 'following the shock of the Dutch elm disease
disaster'.[22]

Ash dieback

Although there have many new tree diseases in Europe since Dutch elm disease, the one that has achieved greatest notoriety is ash dieback disease, caused by the fungus *Hymenoscyphus fraxineus*. The disease is also known as chalara, from its former scientific name *Chalara fraxinea*. This dieback, first spotted in Poland in the early 1990s, was initially thought to be caused by frost and drought but within ten years it became clear that it was caused by a new pathogen, *H. fraxineus*, which originated in east Asia. Its virulence meant that within a decade large areas of Polish ash woods were infected. Unfortunately, ash dieback had spread through much of Europe before the causal agent was identified, which meant that trade in young, infected plants continued throughout most of the first decade of the new century.

The latest research indicates that the planting of infected stock in the UK 'may date back to the early 1990s', which is before 'the first description of the pathogen in mainland Europe in 2006'.[23] However, it was first identified in the UK in 2012 in a Buckinghamshire nursery that had imported stock from the Netherlands. In June 2012 it was also found at a Surrey nursery that had also imported trees from the Netherlands; this nursery had already sold trees to ninety companies and individuals. Nurseries 'were enmeshed with the disease' and its spread was greatly facilitated by international trade in young trees. Some dealers were 'direct importers of plants from mainland EU'; some did not import directly but purchased trees from those who did; others grew some seedlings themselves but imported most of their stock 'from mainland EU because they could be grown much more cheaply there'. It is claimed that some nurseries 'were aware that they had imported ash trees with unusual dieback from which, rather than reporting [the disease], they chose to prune out symptoms before selling'. Indeed, although it is 'probable that the fungus had been carried by wind to the UK from mainland Europe', trade in plants 'greatly sped its introduction'.[24]

The arrival of this virulent form of ash dieback added to the woes of the ash, which had been suffering from another form of dieback since the 1960s. People had been noticing that many mature and maturing ash trees had been suffering from dieback of branches since the 1960s. The Forestry Commission's pathologist Tom Peace, who had worked on Dutch elm disease, described the condition in 1962, after examining trees in Northamptonshire that had been affected by the construction of the M1 in the late 1950s. He thought that it 'was due to moisture stress associated with major variations in soil water content, particularly on heavy soils'. A survey carried out in 1983 confirmed that ash dieback was widespread, affecting on average around 22 per cent of trees, rising to 40 per cent in Northamptonshire. Certainly, when driving around the Midlands during this period the sight of ash trees with many dead or dying limbs was a common one. Interestingly, dieback appeared to be most common in hedgerow trees and was rarely found in woods. Coppice regrowth and young saplings were seldom affected. But what caused this mysterious dieback? Early studies suggested that it was not associated with a virus. There was, however, a clear geographical pattern. Trees in towns and villages were not often affected, but those growing along field boundaries in areas of intensive agriculture frequently suffered from the condition. It seems likely therefore that a range of modern agricultural methods such as deep ploughing, application of fertilizers and pesticides and the then common practice of burning stubble were to blame.[25]

This earlier form of ash dieback is now largely forgotten as, since 2012, virulent chalara dieback has obliterated huge numbers of trees, whether in woods, towns or open countryside. Studies of the early impact of the disease in other European countries made it clear that the prognosis was bleak. The forest pathologist Barnaby Wylder reported that in 2012 the impact in Denmark 'was terrible'. It affected Copenhagen's street trees, and he saw woods 'with recently planted ash more dead than alive' and 'catastrophic dieback in forest stands of mature ash'. But there were some early signs of hope with 'apparently healthy trees monitored since 2007 adjacent to sickly trees'. The UK press used 'apocalyptic terms' to describe the Danish situation, stating that 90 per cent of ash trees were 'killed' rather than the more accurate 'infected'. However, the disease spread rapidly throughout most of the UK over the next ten years and is still very active. Trees most likely to be infected include fresh coppice regrowth and young saplings.[26] I had an acre of thirty-year-old ash trees adjoining a main road felled in 2022 and almost all the ash coppice regrowth and young trees are infected with the disease, while nearby trees that were not felled still appear relatively healthy. Luckily in that small plantation there is much natural regeneration of field maple (*Acer campestre*) and alder, with vigorous regrowth from chestnut coppice, to take the place of the ash.

There is now clear evidence that some older ash trees can resist the disease for several years, and the condition of some trees appears to have improved in the last few years.[27] On the other hand, trees weakened by ash dieback are likely to be more susceptible to attack from other pathogens such as honey fungus (*Armillaria* spp.). Evidence from studies in Czechia and Norfolk indicate that woodland trees growing in damp, rich soils are more susceptible than those growing in dry, exposed conditions. A sensible policy is to fell trees that are near to roads and busy footpaths to reduce the risk of branches falling on people and cars. In other locations it seems reasonable to keep mature trees and to leave maturing coppice uncut. Similarly, large old pollard ash trees should wherever possible be left uncut to reduce the chances of them becoming infected. The retention of as many large old trees as possible will promote the 'production and dispersal of ash seed with resistant genotypes' as part of a long-term programme of recovery.[28]

Chestnut blight

Chestnut blight (*Cryphonectria parasitica*; formerly known as *Endothia parasitica*) originally came from east Asia, where it infects indigenous sweet chestnut species – without, however, causing much damage, because they have co-evolved over a long period. The disease was first recorded in the New York Zoological Park in 1904 and the impact on American chestnuts (*Castanea dentata*) was devastating: by the mid-twentieth century it had spread 'over the entire native range of the American sweet chestnut, from Maine in the north to Georgia in the south, and west to Ohio and Tennessee, and into Ontario and British Columbia in Canada'. The blight killed millions of trees outright and 'has largely eliminated American chestnut as a dominant overstorey tree species'.[29]

The European sweet chestnut (*Castanea sativa*) holds an important place in the cultural and social history of large parts of Italy and France as for centuries its nuts were a staple crop, producing a nutritious flour that could be used to make bread and pasta. Chestnut groves and orchards came to dominate the landscape in many parts of the Apennines, the lower slopes of the Alps and parts of the Massif Central.[30] These large areas of open chestnut woodland meant that when chestnut blight first arrived in Europe from America in 1938 at Genoa, the disease was able to spread fast through the chestnut-growing parts of Italy, soon reaching France, Switzerland, southeast Europe and Turkey. There were probably distinct introductions of the disease to Spain and Portugal and southwest France. It is likely that some of the infected trees were introduced 'to obtain chestnut trees resistant to ink disease . . . which was the most significant chestnut disease before the arrival of chestnut blight'.[31]

Chestnut blight is now common throughout Europe and has recently reached the UK. The import of infected trees was again a key factor. It was first recognized in 2011 at a chestnut orchard in Warwickshire, where around half of three hundred trees supplied by a French nursery in 2007 were infected. Another infection was then found in Sussex, with trees supplied from the same French supplier who, it was discovered, had delivered trees to seven other places. All trees had to be traced and destroyed by burning. Surveys were carried out in the surrounding areas in 2011 and 2016 but no further cases of blight linked to the sites were discovered. A further outbreak was discovered in Devon in 2016 and 'the movement of oak and sweet chestnut material, including plants, logs, bark, branches, foliage and firewood out of, or within, six zones in Devon and Dorset' was prohibited until 2018. At the time of writing (2024) chestnut blight findings were being investigated at 65 sites in England 'to determine the extent of the disease, with a view to remove infected trees where possible'.[32]

The blight affects the branches and trunks of the chestnut tree and causes cankers which can cause its death if they encircle the branch. In America the blight kills the trees, but in Italy and Europe more widely many trees have recovered from the disease due to the natural occurrence of hypovirulence. Hypovirulence happens when *C. parastica* fungus 'is infected by a naturally occurring virus (called dsRNA hypo-virus CHV1) which limits the ability of the pathogen to grow in chestnut bark, or to produce spores'. Hypovirulence was 'first observed in Italy in the 1950s in heavily infected chestnut stands that showed signs of recovery from the disease'. The cankers associated with the hypovirus 'typically have a swollen appearance and are super-ficial or callused', and in many parts of Europe 'chestnut blight cankers display a high prevalence of hypovirus infection'. [33]

This is very good news for the future of the chestnut tree in Europe. But the impact of blight on the production of high-quality nuts was disastrous. Many chestnuts were grafted to improve the quality of the nuts, and they were carefully grown on terraces, even on very steep slopes. In a few places, such as Carro and Valletti in the Val di Vara, examples of traditional, carefully pruned trees associated with small-scale cultivation can still be seen. But in most chestnut-growing areas the combination of chestnut blight and the collapse of the market for chestnuts and flour has led to the abandonment of many chestnut orchards. In many cases the trees felled after the Second World War were converted to chestnut coppice, which grows back vigorously and has a wide variety of commercial uses (illus. 21).

In recent years there have been several attempts to revive high-quality chestnut production, but this has its perils. In 2002 a chestnut grower imported high-quality chestnut graft plants from South Korea to grow in a nut orchard near Cuneo in Piedmont. Very unfortunately this allowed the import of the oriental chestnut gall wasp (*Dryocosmus kuriphilus*). This species only exists as a female and reproduces by parthenogenesis. It is around 2 millimetres long and lays eggs in chestnut buds between May and July that hatch the following year. When the buds burst the feeding larvae produce green or rather attractive pinky-red galls of between 5 and 20 millimetres on the chestnut leaves (illus. 40). This damages tree growth and hence '*D. kuriphilus* is considered one of the most dangerous pests of the genus *Castanea*.' The gall wasp is native to China and was introduced to Japan in 1941, South Korea in 1958 and the USA in 1974. Following its introduction to Italy the oriental gall wasp was classified as a quarantine pest and emergency measures were introduced to prevent its further spread in Europe. However, these have largely been ineffective, and the wasp is now found through most European chestnut-growing areas.[34] Indeed it was identified for the first time in the UK in June 2015 in Farningham Wood, an ancient woodland and nature reserve near Sevenoaks, some street trees in St Albans, north of London, and several other sites in southeast England. The gall wasp spreads fast by flying, being carried in the wind and by the movement of infested chestnut plant material. Luckily a biological control, the parasitoid wasp *Torymus sinensis*, has been successfully used across Europe and this has so far been very effective in reducing the damage caused by the gall wasp.[35]

Fire

Humans have influenced the frequency of wildfires for at least 12,000 years, but it was not until around 5,000 years ago that they became a more important influence on fire regimes at a global scale than 'natural climate variability'. Natural fires are caused by lightning strikes and sometimes volcanic activity, but most fires are caused by humans. Sometimes this is planned burning to increase the quality of land for agriculture or clear woodland; other fires are accidental. A recent international survey estimates that 'there has been a ten-fold increase in the frequency of fire regime change during the last 250 years' compared to earlier periods of the last 12,000 years. This was initially due to agricultural intensification, woodland clearance and other land use changes, and later human-induced climate change associated especially with the Industrial Revolution from the mid-eighteenth century onwards.[36]

In Europe the use of fire for centuries as a form of land management has resulted in a 'complex pattern of land-covers and fire occurrence' that has little if any resemblance to a 'natural fire regime'. The impact of fires has increased following the decline of traditional agricultural practices, including grazing, and the lack of woodland management, which leads to the build-up of large quantities of combustible dead wood. The planting and spread of inflammable species has increased the risk of fires. Controlled use of fire is also one of the traditional techniques used to manage heathland, moorland and grassland in many parts of Europe. The risk of forest fires today varies very much from country to country. Mediterranean countries are particularly prone to fires and Portugal has the largest number. The popularity of plantations of eucalyptus is a key factor, but there are other reasons, including the hot and dry summers following vegetation growth in the mild winters, and extensive land abandonment and the collapse of traditional agriculture, leading to the spread of shrubs and trees over previously cultivated lands.[37]

In Croatia more than three-quarters of forest fires occur in the coastal regions and most take place during July and August. An analysis of the pattern and types of 275 wildfires recorded in the counties of Šibenik-Knin and Split-Dalmatia in 2013 shows the different factors influencing the frequency of fires and the types of land use affected. Around 3,000 hectares of land was burnt, and most were small areas:

three-quarters were patches of less than 5 hectares. There were nine areas larger than 100 ha, including one of 400 ha. There has been extensive coniferous afforestation of parts of Dalmatia over the last 150 years. These plantations had a small number of fires (23), but the area burnt (1,150 ha) was very large. The Aleppo pine is very commonly planted and spreads by natural regeneration. It has been classed as a 'pyrophytic' species because it regenerates well in the burnt areas. Fires encourage the pine cones to open and release the seeds, which fall onto the bare soil, and the burning helps to reduce competition from other plant species. Indeed, this powerful combination of advantages makes the Aleppo pine an invasive species in Dalmatia. Moreover, many areas of Aleppo pine are undermanaged, with large amounts of brush and dead wood accumulating on the forest floor. This stock of dead wood, combined with the live branches, needles and pine cones, which are highly inflammable, makes the risk of severe fire very great indeed.[38]

The geographer Ivan Tekić showed me the effects of fire in the area inland of Šibenik near Vrsno in July 2019. A large fire of the previous year had killed many young pines and junipers, but there was already regrowth of some pines. Some broadleaved trees were also starting to regrow, including *Quercus pubescens* and the manna ash, which had dead main stems but fresh young shoots growing up from the base. At the top of Eagle Hill, pine was regenerating well after a fire and there were young seedlings of juniper. From the hill, looking towards Grebastiča, it was possible to discern patches of the countryside which had been burnt in different years. Areas with bright yellow grasses had been burnt the previous year; the grasses became a duller, greyer colour from the second year after the fire, and patches of regrowing shrubs and trees could be seen. The fire created a complex mosaic of different types of vegetation which slowly began to cover the bare white karst rock (illus. 22).

Land abandonment is one of the most important factors allowing the natural regeneration of trees and shrubs on ground formerly

cultivated and grazed. Another change is the loss of traditional woodland activities such as the collection of firewood. More subtly, many
small-scale agricultural practices such as the raking and gathering of
leaf litter have disappeared from many Mediterranean landscapes. This
is an increasingly forgotten practice, and one which has been dying out
since the 1960s. The leaves were often used for animal bedding and
then, mixed with animal manure, returned to the land. They might also
be burnt in small carefully controlled fires to ensure that the pastures
were not damaged by the fallen leaves. I saw the last traces of this 'managed burning' in the late 1990s in the Lagorara valley, where some local
women were burning carefully raked-up chestnut and oak leaves to
improve the quality of the grazing and make it easier to pick up fallen
chestnuts later in the season. Elderly inhabitants of the Val di Vara frequently describe how the abandoned fields and woodland are *tristi e
brutti* ('sad and ugly') compared to the neat, carefully managed and
beautiful terraces, fields and woods which survived until the 1970s.[39]

The decline of leaf collection and managed burning is also a reason
for the increase in wildfires. Official records and censuses have tended
to ignore this topic and treat the collection of leaf litter as a minor forest
product hardly worth the bother of studying. An oral history study of
the area around Monte Pisano, between Pisa and Lucca, investigated
the implications of the loss of long-standing agricultural practices
such as firewood gathering, leaf litter collection and the picking of
fruits and herbs. The elderly farmers interviewed regarded the current
landscape as 'abandoned, overgrown, and fire prone'. They described
how people used to work together to gather firewood and 'raked leaf
litter and cut ferns for stable bedding, cut brush for kindling, gathered
herbs, berries and mushrooms, and rapidly extinguished small wildfires'. Whereas nowadays paths were often blocked by fallen trees, one
woman remembered that 'you could walk almost everywhere, whether
there was a path or not,' because the ground was raked and cleared of
leaves, twigs and small fallen branches and 'if a tree fell someone
immediately made it into firewood.'[40]

One reason why the importance of leaf raking and associated agricultural practices have until very recently largely been ignored by forest historians is that most of the hard work was undertaken by women and children. A verse recorded in 1861 near Florence described how, 'If I choose to be a peasant woman I will always gather leaf litter, I'll turn as black as tar, I don't want to be a peasant woman.' One woman born in 1935 recalled that her grandmother used a wooden hand rake to gather litter that was then carried back home by donkey. A woman born in 1928 'raked litter every morning before school, beginning when she was about five years old'; children would often work together with school-friends. The leaf litter was carried home in baskets and stored in piles outside the animal sheds ready for use as bedding. The importance of leaf litter collection in preventing the occurrence of large wildfires is only now beginning to be realized. Today, local fire managers are 'deeply worried about the legacies of agropastoral abandonment', which has allowed the accumulation of dry leaves, twigs and small branches intermingled with dried grasses and herbaceous plants. Together with regrowth of shrubs such as *Juniperus communis*, *Cytisus scoparius*, *Genista salzmannii* and young tree saplings, these form a combustible tinderbox for the creation and spread of wildfires. A tinderbox, moreover, that spreads over thousands of hectares and is just waiting for a stray match or spark to initiate a dangerous wildfire.[41]

One way to reduce the risk of uncontrolled fire is to introduce prescribed or managed burning to reduce the amount of dried vegetation in a controlled way. Prescribed burning is the 'planned use of fire to achieve precise and clearly defined objectives.'[42] A well-known example in the UK is the planned burning of heather moors to increase the availability of grouse (*Lagopus lagopus scotica*) for shooting. Regular burning of heather every fifteen years or so means that the heather does not become excessively leggy and creates a mix of burnt and unburnt areas, which can be beneficial for the raising of grouse chicks. It also encourages a higher ratio of younger birds to older birds: 'Regular strip-burning may provide better breeding habitat for red grouse by creating a mosaic of different vegetation heights providing food and cover for chicks.' However, the relationship between such burning, erosion of upland peat bogs and the carbon sequestration potential of such bogs is hotly contested.[43]

Prescribed burning is increasingly being used in southern Europe to help reduce the risk of large wild fires. Some of the earliest experiments were made in Greece in the 1960s, and trials have now been carried out in Portugal, Spain, France and Italy. These experiments are most successful when they link the local knowledge of farmers and woodland managers with fire scientists. In the French Pyrenees, for example, research by the historical geographer Jean-Paul Métailié demonstrated the long history of the use of fire to improve mountain pastures. There were many disputes between peasants and forest authorities, including the famous 'War of the Demoiselles', which started in May 1829 following the introduction of the 1827 Forest Code that restricted rights to graze animals and gather firewood from royal forests. When forest guards tried to stop the pasturing of sheep in the woods, large groups of men 'entirely disguised as women' gathered 'to the call of seashells or cow horns' to reclaim their flocks.[44]

Managed burning of pastures to improve their quality began to be castigated and stamped out by the French National Forest Service when it started a programme of large-scale afforestation in the later nineteenth and early twentieth centuries. Traditional pastoral burning was seen as inimical to the success of the new plantations.[45] As the century progressed, however, land abandonment and the spread of trees began, as in other parts of Europe, to increase the fire risk. In the Pyrénées-Orientales this was linked to 'loss of traditional knowledge and poor burning practices' and in the 1980s a prescribed burning team was established. This included agricultural and forestry experts and firefighters. An important part of the scheme is the 'provision of training to assist communities in restoring sustainable burning practices'. Although prescribed burning could be an important means of reducing fire risk, such schemes are often unpopular because of the impact on air quality and are seen as dangerous by those unfamiliar with the history of land management.[46]

Another reason for the increase in wildfire risk throughout Europe is the greatly reduced collection of firewood. The former importance of firewood in eastern Europe has been examined by the historian Joachim Popek in his study of records held by the Central State Historical Archives in Lviv. He studied the daily life of peasants in the part of Poland formerly controlled by the Habsburgs from 1772 to 1918 and known as the Kingdom of Galicia and Lodomeria (Austrian Galicia for short). These archives hold a remarkable collection of transcripts from peasant hearings on disputes with landlords over the regulation of common rights after the abolition of serfdom by the Habsburg monarchy in 1848. For Galician peasants 'the forest symbolised a source of natural resources to meet the existential necessities of their everyday lives.' The main concerns of the peasant hearings included the collection of firewood and construction wood for cottages and farm buildings. Wood was also used for making agricultural implements such as harrows and ploughs, well casings, animal feeding troughs, many utensils and fences. In addition, as across Europe, forests were used to graze cattle and swine, provide shelter for animals in hot weather and heavy rain, and as a place to collect fruits, mushrooms, acorns, beechnuts, herbs and leaf litter.[47]

The importance of firewood is indicated by the complexity of the associated regulations. The tree species used were mainly coniferous or broadleaved softwoods such as spruce, fir, larch, pine, poplar or alder. Firewood was used throughout the year for cooking and providing hot water for washing and in the winter was essential for heating the main room in the house. Firewood was frequently divided 'between "light gathering" (*drobnazbiórka*) and "heavy gathering" (*grubszazbiórka*).' Light gathering consisted of otherwise largely valueless leftovers from forestry operations and was further divided into four main categories. First was *wierzchy*, which included the tops and branches of felled trees. These had no timber value as they were full of knots and no more than 2 inches thick. Second was *leżanina*, which was coarse woody debris and thick pieces of bark. Third was large and small standing dead trees. These were often collected in the winter when snow covering the ground made it easier to move the wood around. This category included young dead trees known as *podsuszki*: 'After cutting down such a tree or digging it out with roots, the peasants were required to cover the hole and level it with forest litter.' The fourth category of light firewood was tree stumps (*karpy or pniaki*), sometimes from freshly felled trees but more usually older rootstocks. This was very hard work indeed: 'they dug out older and decayed *karpy* (rootstocks) with their main root systems by prising them out with a wooden stick.' Alternatively, they partially excavated the root system with hoes and cut off the usable firewood with an axe.[48]

Heavy gathering of firewood described gathering larger branches and fallen trees damaged while felling or in storms. It also included large old dead trees that could not be used for construction timber. This type of firewood required an axe to cut up the thick branches and the payment of a fee to the landowner was usually required. Where there was a good relationship between the estate administration and the serfs there was sometimes a fixed tariff for the amount of firewood provided, with the manor allocating 'a given number of trees of particular species to each unit of peasant farmland'. In the village of Przegonina, for example, the peasant farmers 'received two beeches and two firs each in the autumn of every year'. This meant that the farmers had no need to worry about their firewood supplies. Analysis of the archives showed that after 1848 those estates with large areas of forest 'had clear, specific regulations for the gathering of wood, use of tools, conduct in and movement around the forest'. The firewood collection season was usually after harvest from September to December.

The head forester and forest wardens would usually supervise the collection of firewood. 'On set days, in the morning hours, the peasants gathered at the forest tollgates' and showed their written permits, which listed 'the peasant's given name and family name, place of residence, house number, signature from the manor farm administrator or the owner of the estate, and the authorized number of cartloads'. In exchange for the permit the peasant was given 'the so-called "forest stamp" (*znaczeklasowy*), [which] is a wooden plaque or piece of paper containing an impression of the landlord's seal'. Waggons were searched to ensure that only authorized wood was collected and illegal saws or axes had not been used. These rules were strictly enforced: 'Any farmer caught committing such an offence (or cutting down a healthy tree) was ordered to unload the entire cargo and go back home with an empty cart.'[49]

Although firewood remains a relatively cheap, and fashionable, form of heating in many rural and some urban areas, it has declined in importance. Its cultural significance was demonstrated by Natalia Ginzburg in her short story 'Winter in Abruzzi', written in Rome in 1944, which tells of her wartime exile. Fires were usually burning in farmhouse kitchens but these varied depending on the prosperity of the family. There were

great fires of oak logs, fires of branches and leaves, fires of twigs picked up one by one in the street. It was easier to tell the rich from the poor by looking at the fires they burnt than by looking at the people themselves, or at their clothes and shoes which were almost all more or less the same.

The contemporary Tuscan crime writer Valerio Varesi in *The Dark Valley* (2005) reveals the decline in interest in wood fuel. He sets his novel in the fictional town of Montelupo, which is 'going to the dogs. There's no one left who's willing to clear the ditches, to attend to the drains or to look after the woodland. Instead of going to gather firewood, they switch on the gas.'[50]

5

FELLING TREES

I t is relatively rare for trees in Europe not to be felled at some stage in their life. Most are felled to produce firewood, timber or other wood products. Trees remain of great importance for construction timber and in some countries, such as Norway and Sweden, many houses are still built almost entirely of wood. Different styles of vernacular architecture, from Alpine chalets to the half-timbered cottages and houses of parts of France, England and Germany, help to define cultural landscapes. The use of wood for construction is now seen to be of enormous importance in helping to mitigate climate change. It does this through the storage of carbon sequestered by trees in buildings and by replacing other construction materials, such as steel and concrete, and thus reducing carbon emissions.[1]

Some trees may be retained as dead, standing trees for conservation reasons, but if there is public access to woodland or the trees are near roads or adjoining footpaths, these are likely to be felled at some stage for safety. Falling trees are a real danger to passing traffic, people, power lines and buildings.[2] But some woodland owners do allow trees to grow and die without felling them. Myles Hildyard of Flintham Hall, Nottinghamshire, told me in 1978 that his policy was to allow mature trees in the old landscape plantings around his house to grow old, die and remain in place for landscape and conservation reasons.[3] In contrast, woods adjoining the Fosse Way (A46), which bisected his estate, had to be felled for safety reasons. He was a keen supporter of the National Trust and his treatment of trees accorded with the views of James Lees-Milne, the diarist and influential Historic Buildings Secretary of the National Trust, who wrote in October 1946, 'The truth is that all aesthetes hate any trees being felled whereas counter-aesthetes love felling as many as they can.'[4] The context here is that the architect Harry Goodhart-Rendel, who had given his Surrey mansion Hatchlands to the National Trust in 1945, was distressed to discover that the National Trust land agents had 'decided to cut down some trees in his park contrary to his wishes'. Lees-Milne was 'determined that this sort of thing must be put a stop to'. His view was strengthened when he visited Hatchlands on 4 May 1947 with Goodhart-Rendel, who was 'very upset by the N.T.'s ridiculous injunctions to cut down the trees at Fuller's Farm'. Lees-Milne thought it 'intolerable of the mangel-wurzels', his nickname for the land agents, 'to dare dictate to' Goodhart-Rendel.[5]

Timber felling is sometimes seen as a sign of weakness. Timber may be retained as something to be felled during times of financial crisis, the felling associated by observers with bankruptcy. The wealthy Gertrude Savile noted in her diary on 10 August 1727 that Sir Theodore Janson, formerly a director of the South Sea Company, had pulled down his 'finly scituated' old house at Wimbledon. She classed him to be an 'Upstart if from no other reason he pull'd down a fine stately house to build with the materialls a nasty, naked modern one'. She reported that there remained 'fine Woods and plantations which he [Janson] was cutting all down. But he was Lop'd himself before he had accomplish'd it. So that there are yet great beauty's remain, tho' in Ruins which, I don't know, may yet add to the beauty; it gives them, a melancholy, solemn look.'[6] William Wordsworth used his *A Guide through the District of the Lakes* (1810–35) to deplore the unthinking felling of trees. He describes with horror an example of large-scale clear-felling: 'The axe has here indiscriminately levelled a rich wood of birches and oaks, that divided this favoured spot into a hundred pictures'. These 'beautiful woods' were felled and 'scenes, that might formerly have been compared to an inexhaustible volume, are now spread before the eye in a single sheet' that can be 'perused in a moment'.[7] Dr Astrov in Anton Chekhov's *Uncle Vanya* (1899) carefully mapped woodland loss:

> (*pointing to the map*) Now look here! It's a picture of our district as it was fifty years ago. The dark and light green stands for forests: half of the whole area . . . Let us pass to the third map – a map of the district as it is at present . . . There is green only here and there . . . it's a picture of gradual and unmistakable degeneration.[8]

Thomas Gainsborough depicted the woodman as an honest labourer sheltering with his dog under a decayed pollard (illus. 30), with a bundle of cut branches. He thought this painting, finished in 1787, to be his best and it 'was shown to George III at Buckingham House' shortly before Gainsborough died.[9] But for many commentators, the woodman was the villain of the piece. The nineteenth-century Church of England clergyman and diarist Francis Kilvert was a sensitive observer of trees who carefully related his personal experiences and ideas about particular species. While strolling on Sunday, 23 April 1876 at 'calm and serene' Monnington-on-Wye, he noted that he loved 'to walk up the great avenue, as up a vast and solemn Cathedral aisle, while the wind sighs through the branches of tall sombre Scotch firs overhead and makes mournful music'. A nearby 'silver birch droops and waves her long dusky tresses as a maiden with delicate white limbs and slender arms lets down her long hair and combs it to the curve of her beautiful knees'. A few years earlier, while curate at Clyro, Radnorshire, he reported that 'Some barbarian – a dissenter no doubt – probably a Baptist, has cut down the beautiful silver birches on the Little Mountain near Cefn y Fedwas.' He was even more upset on 3 February 1872 when he discovered that 'the beautiful Cwmgwanon

30 Peter Simon, after Thomas Gainsborough, *The Woodman*, 1790, etching.

woods are to be felled.' The local cottagers told him that the felling was to pay for the gambling debts of the landowner Francis de Winton of Maesllwch Castle. The trees were 'doomed' and had been 'measured and numbered' and soon 'the axe will be heard in the sacred dingle.' Just over a month later on 4 March he walked to the dingle and recorded that 'Cwmgwanon Wood is being murdered.' He 'looked sadly down into the hollow, numbers of my old friends of seven years standing lay below on both banks of the brook prostrate and mutilated, a mournful scene of havoc, the road almost impassable for the limbs of the fallen giants'. Later in the month he described how the 'timber carriages which are hauling away the fallen giants, ash and beech' had 'fearfully cut up' the meadows. 'The shouts of the timber hauliers were ringing hollow and echoing through the wasted murdered dingle. My beautiful favourite Cwm is devastated and laid waste.'[10]

Every so often the felling of an individual tree raises a public outcry. This is especially true of trees that have a strong cultural value such as ancient trees, named trees and trees that have been famous because of their literary or film connotations. In Britain, if the trees are in Conservation Areas, mainly in villages and towns, or have Tree Preservation Orders, special permission has to be gained from a local authority to fell or lop a tree. Throughout the country felling licences should be gained for the felling of trees in woods, plantations and hedgerows. Trees in most commercial plantations are destined to be felled. After the trees have reached 'pole stage' the crop is likely to be thinned to allow the surviving trees to grow more quickly and productively. The final crop may be clear felled, with the whole of the compartment or stand of the wood felled at once, before regeneration takes place. Alternatively, forms of continuous-cover silviculture involving the selective removal of trees may be used. In a highly populated country it is difficult to keep the felling of trees secret. The notorious felling of the Sycamore Gap tree in Northumberland in 2023 took place under cover of darkness. Its isolation and situation alongside Hadrian's Wall in Northumberland, together with its appearance in the film *Robin Hood: Prince of Thieves* (1991), had made it an internationally known celebrity. It was perhaps this fame which made it a target for assailants. Happily, sycamores are well known for their resilience and the expected regrowth of this tree soon materialized in summer 2024.

Binsey poplars

On 13 March 1879 the poet Gerard Manley Hopkins went for a walk from Oxford, where he was a curate at St Aloysius' Church, to Godstow by the River Isis (Thames) and was horrified to see 'that the aspens that lined the river are everyone felled'.[11] The same day he wrote the poem 'Binsey Poplars', the opening lines of which are among the most famous evocations of the effect of tree felling on a human imagination:[12]

> My aspens dear, whose airy cages quelled,
> Quelled or quenched in leaves the leaping sun,
> All felled, felled, are all felled.

The immediate reason Hopkins was impelled to write this poem was his discovery that some of his favourite trees had just been felled. He was horrified to witness the destruction of trees he had first known as an undergraduate around fifteen years earlier. But his love of trees and hatred of felling had a long history. As a boy he had 'a love of climbing to the top of tall trees, where he would remain balanced for hours, gazing with wonder at the landscape'. This also allowed him to keep a distance from his family. 'The tall elm' at their family home in Oak Hill Park, Hampstead, 'was his usual platform' but on visits to his grandparents at Blunt House, Croydon, which had a 'well-timbered park', he 'practised his climbing skills' on 'two cedars of Lebanon and a large old beech' growing on the lawn near the house. As an undergraduate at Balliol College, Oxford he was furious when a large beech tree was felled 'merely to make more light for the rooms of the Fellows'.[13]

Hopkins, a keen amateur artist, paid very close attention to the pattern of foliage and blossom and seasonal changes. When walking near Oxford on 3 May 1866 he noted that some 'oaks are out in small leaf. Ashes not out . . . Elms in small leaf, with more or less opacity.' At Skinner's Weir people 'were peeling osiers which gave out a sweet smell'. The following day he was 'Alone in Powder Hill wood. Elms far off have that flaky look now but nearer the web of springing green with long curls moulds off the skeleton of the branches.' In June 1866 he noted that 'It is the tufts of bloom on Spanish chestnuts crowning the round tufts in which the leaves are thrown, which make those wavy concentric outlines this tree has at twilight.'[14]

While Hopkins was at Manresa, the Jesuit training school at Roehampton Park, he revelled in the fine trees. When picking mulberries in the summer of 1869 he noted:

When you climbed to the top of the tree and came out the
sky looked as if you could touch it and it was as if you were
in a world made up of three colours, the green of the leaves
lit through by the sun, the blue of the sky, and the grey
blaze of their upper sides against it.

But not long after they had finished picking the fruit 'a very great gale . . . broke the mulberry tree near the farm by the ground and struck half the cedar in St Aloysius's walk into the rye-grass field . . . It frightened one to go among the trees.'[15] A few years later when living at Stonyhurst College in Lancashire he was saddened by the felling of an ash tree in the garden: 'It was lopped first: I heard the sound and looking out and seeing it maimed there came at that moment a great pang and I wished to die and not to see the inscapes of the world destroyed any more.'[16]

Robert Martin, Gerard Manley Hopkins's biographer, argues that his 'sorrow at the destruction of trees was to remain as fresh at the end of his life as it was when he was an undergraduate, for it seemed to symbolise all that he hated about the eradication of the past'. The felling of the poplars next to the Isis at Binsey desolated Hopkins: 'The trees that he had known and loved since he was a brand-new freshman were gone, and with them had gone his youth.'[17] The poem has strong common themes with William Cowper's 'The Poplar-Field', written in 1784 and published in 1800:

> The poplars are fell'd, farewell to the shade/ And the whispering sound of the cool colonnade.' The poet 'last took a view/ Of my favourite field and bank where they grew' twelve years ago, 'And now in the grass behold they are laid/ And the tree is my seat that once lent me a shade.

Cowper's influential poem demonstrates many similar sentiments about the passing of time and the loss of youth.[18] Hopkins's familiarity with Cowper's poem will have reinforced his dark impressions when he came across his felled poplars by chance.

There was, moreover, another reason for tree felling to be uppermost in his mind in March 1879. His father, Manley Hopkins, was deeply involved in a campaign to stop the proposed felling of one side of an avenue of trees along Well Walk, near the family home in Hampstead. As part of his campaign Manley Hopkins wrote a poem, 'The Old Trees', which was published anonymously in the *Hampstead and Highgate Express* on 28 December 1878. The poem looks back fifty years when a girl and a boy 'neath these limes . . . sat, in joy' and the 'nightingale sang on the Heath', and then later 'Under these trees, in married love/ We paced so oft when day was o'er.' The poem emphasizes the spread of London, the loss of trees and how 'scenes that nursed our love have changed', but asks 'for a little space/ Ere *these* trees fall beside their race'. It was part of a much larger campaign, including letters to *The Times*, which emphasized the association between the avenue of lime trees and figures such as Dr Johnson, John Constable,

Leigh Hunt and John Keats, 'whose favourite seat was at its eastern extremity'. The protest was very effective: the *Hampstead and Highgate Express* reported on 18 January 1879 that the trees in Well Walk would be spared (illus. 39). Gerard Manley Hopkins would have heard of the success of this campaign from his family, and so would have been particularly shocked by the felling of the poplars he knew so well at Binsey.[19]

A question remains concerning Hopkins's naming of the poplars as aspens. He was a very acute observer of trees and would have known that the aspen (*Populus tremula*) is characterized and named after its small leaves, which tremble in the slightest breeze. But it is unlikely that the felled trees were aspens, which spread vigorously through suckering. The tree is often found in woodland and individual clones in eastern England can reach 5–10 acres in area.[20] The suckering form of the aspen makes it very tenacious and there are no aspens found at Binsey today. Loudon named this species the 'trembling-leaved Poplar, or Aspen', noting that it had little commercial value and 'As fuel, the wood is of feeble quality'. He stresses that the 'great drawback to the tree . . . is the number of suckers which it throws up; and which, if not eaten down by cattle or mown, would soon turn a whole country into an aspen forest'.[21] Indeed, the closely related American aspen *Populus tremuloides* forms very extensive clones, and one named Pando, in the Fishlake National Park, Utah, spreads over 106 acres, has over 40,000 trees and is thought to be 'the largest, most dense organism ever found'.[22]

Another contender could be the native black poplar (*Populus nigra*), but there is no historical evidence of this rare species growing in the vicinity. It is much more likely that the poplars were one of the species of hybrid poplar introduced from Italy and Canada in the late eighteenth century and the early nineteenth. These different species of poplar have a complex history of hybridization and introduction. The black Italian poplar (*Populus* × *canadensis* var. *serotina*), which is one of the commonest, is probably a cross between a European

(*Populus nigra*) and a Canadian or American poplar (*Populus deltoides*) which may have originated in France in the eighteenth century. The naming and history of these trees was mysterious and contentious in the early nineteenth century. The usually precise Loudon was rather confused. He thought that the black Italian poplar (*Populus monilifera*) 'was introduced into England in 1772, from Canada', but it could also have originated in Georgia. He noted that the names 'Swiss poplar, and black Italian poplar, allude to the tree being very abundant in Switzerland and the north of Italy'. However, he suggests that 'Notwithstanding [the] evidence in favour of its being a native of North America, we think ... that P. monilifera may have been originated in Italy or Switzerland, and carried to North America; and, if so, this will readily account for the English name of black Italian.'[23]

The history of the tree's introduction was also a source of controversy. It took a while for the value of the tree to be recognized by nurserymen and planters. The Scottish nurseryman Archibald Dickson, from Hassendean near Hawick, popularized the tree after obtaining one 'from a gentleman in that part of Scotland, who had it sent from America by a son then resident there'. Mr Dickson 'travelled for the firm through most of the northern districts of England; and, having a high opinion of this poplar, of which he had been the first to procure stock of plants, he recommended it everywhere.'[24] Initially there was considerable confusion between the black Italian poplar and the Lombardy poplar. The Lombardy poplar, with its very characteristic fastigiate shape, introduced in the mid-eighteenth century, was soon a popular landscaping tree, but produced timber of little value. William Pontey thought that 'the very extraordinary encomiums which used to be lavished' on the Lombardy poplar were due to the confusion between the two species: 'The Black Italian is very different from the Lombardy in form, as it almost uniformly rises with a light, but regular conic head, being so hardy as generally to preserve the same leader, from the ground to an immense height; in consequence its stem is remarkable straight.' It is 'an astonishingly quick grower on every sort of soil that may be called tolerable'.[25] Since the opening years of the nineteenth century 'the black Italian poplar has ... been far more extensively planted than any other species of variety of the genus'.[26]

The Binsey poplars had been planted to be felled. Both the pleasure and the anguish they gave to Hopkins was, from the perspective of their owner, merely incidental. They were probably black Italian poplars, which by the mid-nineteenth century were frequently planted because of their rapid growth, especially in well-watered situations. They were, and still are, frequently planted in non-woodland sites in rows and relatively small groups to encourage rapid growth, which means that their felling is more visible to passers-by than that of other species growing within woodlands. Their rapid growth also means that they are likely to be felled within a human lifetime. Poplars had a range of uses including 'railway brake-blocks ... beds of wagons and wheel-barrows, for second-class spade handles' and 'blocks for polishing plate glass in the course of its manufacture'.[27] Robert Martin argues that the felling of the Binsey poplars was 'made even more poignant because the trees had been cut down for use as brake blocks, or "shoes" for the locomotives of the Great Western Railway', a company that Hopkins saw as 'the arch-enemy of the beauty of the Thames valley landscape'.[28]

Gladstone 'axing questions'

In the same year that Gerard Manley Hopkins was lamenting the fate of the Binsey poplars, William Ewart Gladstone (1809–1898) was at the peak of his tree-felling enthusiasm. He became so well known for this that he was presented with many axe-related gifts including a silver axe-shaped pencil given to him by the Princess of Wales 'for axing questions'.[29] Gladstone started his thorough, terse diary as a schoolboy at Eton: his first entry for 16 July 1825 begins 'Read Ovid ...'[30] His concise, documentary diaries record the thousands of letters he wrote, people he met and, from the 1850s to the 1890s, the numbers of trees he felled. The historian Peter Sewter has analysed the diaries and shows that Gladstone's tree fellings took off in the 1860s, were most prolific in the 1870s and still significant in the 1880s. In the five-year period from 1876 to 1880 Gladstone notes 'axe work' by him and members of his family taking place on 292 days, an average of 58 days a year.[31] How did Gladstone, the foremost Liberal politician of the Victorian age, who was four times Chancellor of the Exchequer and four times prime minister, fit tree felling into his extraordinarily busy life?

His country estate was Hawarden Castle in Flintshire, Wales, which had been his wife Catherine's property. The Georgian house was transformed into a picturesque Gothic mansion in the early nineteenth century. In the grounds the ruins of the medieval castle were retained as a focal point. The village is to the northwest of Hawarden Castle and the park to the south. There had been, as with many landed estates, considerable tree planting in the park and in the woodlands throughout the eighteenth and nineteenth centuries, for example at Bilberry Hill. The landscape gardener William Sawrey Gilpin, who was strongly influenced by the ideas of Uvedale Price, advised on the layout of the grounds in 1830, including the enhancement of the views from the main windows of the house over the parkland. There is a nineteenth-century iron deer fence protecting the woodland planting to its south.[32] By the 1860s there were trees of many ages growing in the gardens, park and woods at Hawarden. Plantations, groves and clumps of parkland trees established following the rebuilding of the house in the 1830s would have needed thinning, as would rapidly growing trees encroaching on views and vistas. Gladstone would walk around the estate with Catherine, who had been born at Hawarden in 1812, to help select trees that should be felled and also decide where to plant new ones.

The flourishing of Gladstone's interest in tree felling coincides with time in opposition (1855–9) following his first period as Chancellor of the Exchequer. Gladstone's diaries show that his Christian faith allowed for few idle moments. When staying at Hawarden he attended church in the village daily and then devoted his mornings to letters, reading, meetings and writing, 'afternoons to walking or woodcraft, and evenings to family entertainment'.[33] Following an early reference to felling three beeches on 29 October 1852, the first detailed mention of tree management at Hawarden is on Saturday, 31 July 1858. In the morning after church, he was writing letters and reviewing his old poetry manuscripts, in the evening he was reading Hume and Cicero. But he 'Spent the afternoon in woodcutting & the like about the old Castle: my first lesson.' The following Monday he had 'A wood cutting afternoon about the lawn & old Castle'. Roy Jenkins notes that there 'were another seven arboreal assaults during that August'. On Thursday, 5 August, Gladstone remained at Hawarden 'Tree-cutting while the party went to Mold for SPG': he preferred tree felling to attending a meeting of the Society for the Propagation of the Gospel with a family party including Bishop Wilberforce. On 24 August he noted that 'We have now also ordinarily a family song or dance after dinner. Wood cutting all afternoon.' He was also reading Rousseau, Mirabeau and Hume.[34]

By the 1860s tree felling was a regular part of his daily routine. He enjoyed felling trees with his sons William and Harry. On 23 December 1867, after writing letters, 'Willy & I felled a good tree . . . A splinter struck my eye, causing some inflammation: and in the evening Dr Moffatt came & sent me to bed.' A couple of days later on Christmas Day, 'Dr M made a very good report. There is however a very slight indentation, close to the pupil.' But he made a rapid recovery and by 27 December he felled a tree in the company of Harry: 'Today a tree we were cutting fell with Harry in it. He shewed perfect courage & by God's mercy was not hurt.' Despite Harry's close shave, the next day they carried on as normal: 'Whist & music in evg. Woodcutting in aft. Read Baker's Abyssinia.' The confirmation of his appointment as prime minister in 1868 arrived as he was tree felling. On 2 December, after writing letters, he recorded 'A little revision of Homer. Read Swift.

We cut down an ash. Gen. Grey arrived at 5¼: with H.M.'s letter.'[35] Queen Victoria had sent a telegram asking him to form a government. Gladstone read the telegram and continued tree felling. A few minutes later, he stopped, looked up and 'with deep earnestness in his voice, and great intensity in his face, exclaimed: "My mission is to pacify Ireland". He then resumed his task, and never said another word till the tree was down.'[36] He had an audience with the Queen at Windsor the following day.

Becoming prime minister did nothing to assail his enthusiasm for tree felling and much was made of Gladstone's skill with the axe

31 Francis Carruthers Gould, *Plenty of Work on Hand*, c. 1876–7, drawing.

32 George
Cruikshank,
The Upas Tree,
c. 1842, etching.

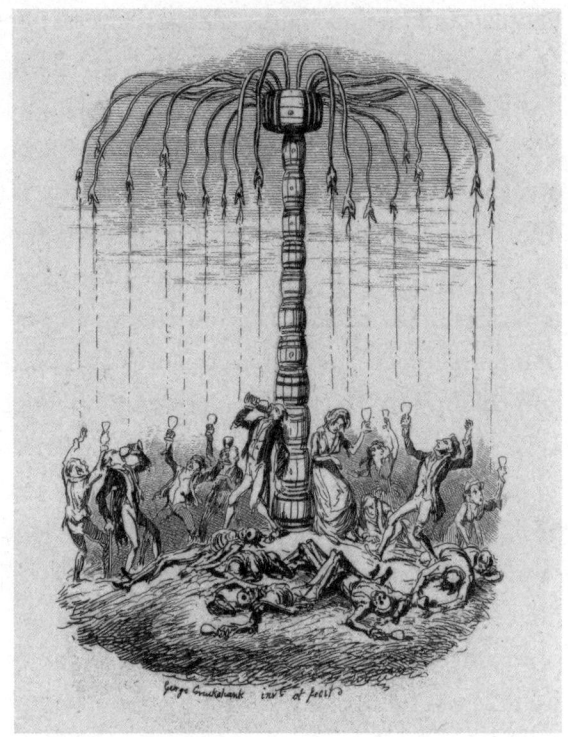

by political cartoonists and commentators. Francis Gould's cartoon
of circa 1876–7 shows Gladstone in shirtsleeves and holding a large
axe pondering the felling of a large, old, hollow, pollarded tree with
a bearded face labelled 'Turkish Atrocities'. In the background similar
pollarded trees labelled 'Drink', 'Lords' and 'Welsh Church' await
their fate. He was celebrated or castigated for his success or failure in
rooting out and cutting down injustices (illus. 31).

One metaphor which took hold was the felling of the tropical
upas tree (*Antiaris toxicaria*) from Java. Knowledge of the devastating effect of its poisonous sap had been popularized by Erasmus
Darwin's poem *The Loves of the Plants* (1789) and it was used in the
nineteenth century to illustrate evils that should be eradicated.
George Cruikshank's cartoon *The Upas Tree* (c. 1842) depicts a bizarre
tree whose trunk consists of barrels of liquor. Its drooping and weep-
ing branches exude poisonous liquor, which dancing and drunken

people catch in their raised glasses. Several people have collapsed, and skeletons litter the ground (illus. 32). Temperance reform was a key element of Liberal policies in the late nineteenth century.[37] Nathaniel Smyth, secretary of a Liverpool temperance society, argued at a conference in Philadelphia in 1876, 'The time is coming when public opinion will be so ripe in England that our great Gladstone, who is very fond of hewing down trees, will have to take up a very different axe from that he has been wielding, and hew down the upas-tree of intemperance.'[38] The upas tree was also used to represent greed and corruption. Joseph Keppler's cartoon *The Deadly Upas Tree of Wall Street*, published in the influential American magazine *Puck* on 30 August 1882, depicts a large, leaning upas tree with Albany, the New York state capital, in the background to the right, and the Capitol in Washington, DC, on the left. The face of the railway magnate Jay Gould, famous for bribery and corruption, is outlined by branches in the centre of the tree. The tree's leaves are gold dollars and its pink flowers list different types of bribe. Famous financiers, politicians and industrialists, such as former president Ulysses S. Grant, who is holding a document labelled 'Black Friday', lie among skeletons under the tree (illus. 41).

In an election campaign speech on the Irish Question at Wigan on 23 October 1868, Gladstone argued that Liberals aim at the destruction of the system of ascendancy in Ireland, which is like 'Some tall tree of noxious growth, lifting its head to Heaven and darkening and poisoning the atmosphere of the land so far as its shadow can extend'. He hoped the day had come where 'the axe has been laid to the root [Loud Cheers]. It is deeply cut round and round. It nods and quivers from top to base [Cheers]. There lacks, gentlemen, but one stroke more – the stroke of these Elections [Loud Cheers]. It will then, once for all, topple to its fall ...'[39] The editorial of the Liberal-leaning magazine *Fun* for 12 February 1870 thought there was much work for the prime minister at the start of the parliamentary session, even for one who 'finds his relaxation in felling timber'. This was especially true of 'an ugly bulk that needs to be hewed down in a neighbouring island'. Gladstone still had a lot to do to 'bring down the Upas tree of Wrong, which was planted so long ago in a fertile soil that is as deeply rooted as Bigotry', and he needed 'a keen edge to his axe!'[40]

Initially Gladstone's tree felling was mainly a private affair, an integral part of family life at Hawarden. By the 1870s his enthusiasm and skill with the axe was so well known that he regularly felled trees when making visits across the country. This was in addition, of course, to his frequent planting of memorial trees. He visited Nottingham in May 1875, combining public and private roles: he was executor of the estate of the 5th Duke of Newcastle, whose vast holdings included Nottingham Castle and its Park, and Clumber Park to the north of Sherwood Forest. Nottingham Castle, which had been attacked and burnt during the intense 1831 Reform Bill riots, was being converted into the Midland Counties Art Museum, opened in 1878. The Park was developed as a housing estate for wealthy local residents. On 10 May 1875 Gladstone 'Met the Museum Committee at luncheon. Went over the Castle grounds.' He 'took another view of the "Monk's Holes"', thirteenth-century medieval caves cut into the sandstone in Notting-ham Park as a hermitage for monks from Lenton Priory, then he 'Dined with the Yeomanry'.

The following morning, 'Before breakfast I cut down a Siberian Elm in the Park, bad axe but soft tree 4f. 6 ord. measure.' It took him fifty minutes. The tree felled is unlikely to have been a Siberian elm (*Ulmus pumila*) as this was introduced to Britain about 1860.[41] It could have been a *Zelkova*, the Caucasian elm, introduced in 1760. We will never know, although a plaque celebrating its felling remains in the Park at Nottingham today. This is a rare example of a memorial to the felling of a tree rather than the tree itself, although the already weak link is attenuated by the fact that the memorial is a modern re-placement; the original was removed to protect it from vandalism.[42] Gladstone then went on to visit the museum, the mayor's lace factory and the nearby sixteenth-century Wollaton Hall, with its 'remarkably

33 Samuel E. Poulton, *The Late Rt. Hon. W. E. Gladstone
Tree Felling at Hawarden*, c. 1888, photograph.

fine exterior', before taking a train to Mansfield and a carriage to
Clumber Park through Sherwood Forest for 10 miles. He thought,
perhaps influenced by his visit to the medieval hermitage the day
before, that the country was 'just such as Lancelot might have traversed
when bringing Guinevere.'[43]

On a visit to the Duke of Bedford at Woburn Abbey on 22 October
1878 he toured the local town with the duke and saw the school
and trees: 'Also I cut down a Deodara [*sic*] assisted by the Duke.' On
3 November 1879 while visiting Wellington College, the public school
in Berkshire founded by Queen Victoria in 1859, he 'addressed the boys
for ab. 40 min. Then cut down three small birches.' The fame of his
fellings meant that when large parties of visitors arrived at Hawarden,
they often expected to see him in action. On Saturday, 4 August 1877,
after six hours of letter writing, 'A party of 1400 came from Bolton! We
were nearly killed with kindness. I began with W[illy] the cutting of

a tree; and had to speak to them, but not on politics.' It is recorded that 'the very splinters which flew from this axe were picked up and treasured as relics.' Gladstone's extensive experience of tree felling meant that he had some favourites. He reported in 1878 that 'Oak, though very hard is not a bad tree to cut, for the grain breaks off easily and does not cling to the axe.' Beech and ash are 'the two most difficult to fell of our English trees' as they bend to the axe, and ash in addition 'is subject to fracture'. He found the Spanish Chestnut 'the pleasantest to fell' and the 'Yew is the most horrible to cut of all forest trees'.[44]

By the 1880s the pace of his tree felling started to decline, and he recognized at the age of 72 that his strength was not what it was. On 22 September 1881 he 'Ventured on a little woodcutting – But I am no longer equal to the true woodmans work' and the following month 'felled a beech with Willy. But *non sum qualis eram*.' However, he was still felling four years later and on 28 May 1885 he 'Cut down a holly, which impeded view. One of my last falls?' He did some light work on 20 November 1889 with his son Herbert, 'I play second fiddle,' and later that month was pleased to see that his grandson Will 'was much interested in seeing the axe and bruited it abroad'.[45] After Gladstone's death in 1898 a memorial postcard entitled 'The Late Rt. Hon. W. E. Gladstone Tree felling at Hawarden' was produced, based on a carefully posed photograph (*c.* 1888) by Samuel Poulton (illus. 33). It shows a large, felled tree with Hawarden Castle in the background. Catherine Gladstone stands centrally in profile in front of children in a donkey carriage; Gladstone rests against the felled tree with an axe over his shoulder, his hat capping the trunk upon which sits his son Willy.

Charlotte Mew and Thomas Hardy

The poet Charlotte Mew (1869–1928) was brought up in Bloomsbury, London, and was very sensitive to tree felling in London streets, squares and parks. Her essay 'Men and Trees', published in February 1913, opens with the declaration,

The London trees are all prisoners of men, some unreason-
ably mutilated like the lopped crowd in Greenwich Park,
while, now and then, there is a wholesale massacre such as
that of the seven hundred in Kensington Gardens, which
took place, no one knows why, some thirty years ago.

In addition to her criticism of the pollarding of trees at Greenwich,
Mew here refers to the controversial felling of many trees in
Kensington Gardens in the autumn of 1880. A correspondent to *The
Times*, 'Chancing to walk past the end of Porchester terrace, and
looking over the park and gardens opposite', could 'scarcely believe
my eyesight'. There were 'piles of timber trees for hundreds of yards
towards the Round Pond'. The 'formerly beautiful glade was like a
wharfinger's timber yard, with the addition of the ground being full
of holes where the roots of the trees had been'. These 'magnificent
timber trees', including 'elm, beech and horse-chestnut ... cannot be
grown again under 100 years'.[46]

This destruction was also lamented by the naturalist W. H.
Hudson, who recalled visiting the ancient rookery in Kensington
Gardens on his first visit to London: 'What wonderfully tall trees
were these where the rookery was placed! It was like a wood where the
trees were self-planted, and grew close together in charming disorder,
reaching a height of about one hundred feet or more.' He was aston-
ished that 'so noble a transcript of wild nature as this tall wood' could
exist 'so near the heart of the metropolis, surrounded on all sides by
miles of brick and mortar and innumerable smoking chimneys'. Most
of the trees were sound and not dangerous. Hudson 'examined many
of the finest boles, seventy and eighty feet long, and could detect no
rotten spot in them, nor at the roots'. The only reason he could discover for the felling was that grass did not grow well in the shade and
when the leaves fell 'the ground was always sloppy and dirty under
the trees, so that no person could walk in that part of the grounds
without soiling his boots.' The noisy and expensive collection of fallen

leaves to allow grass to grow remains a bone of contention for park managers, visitors and local residents to this day. But this seemed a paltry excuse to Hudson. The loss of hundreds of trees and the rookery meant that he could 'feel nothing but horror at the thought of the unspeakable barbarity the park authorities were guilty of in destroying this noble grove.'[47]

Mew lived for most of her life in Bloomsbury and was very concerned that in this, 'her own wooded neighbourhood', the trees' fate is that 'one after another falls; progress pulls down the old spacious shabby houses and puts up flats . . . the green vanishes: even tomorrow one may miss the familiar plane of yesterday, and the birds go with the trees.' She was horrified by the mutilation of trees associated with felling, reporting, 'I really can't bear to see a tree cut down – a big tree: it's a sort of sacrilege.' She described 'a tree outside the British Museum they were felling last week, with all the instruments of butchery, the axe and the rope and the saw, and the clearing round it like a scaffold; it went on for days and I didn't altogether care for it.'[48] In a short poem, 'Domus Caedet Arborem', she personified the Bloomsbury houses and the nearby London planes (*Platanus* × *hispanica*). They lived close together but the buildings, and the rapidly developing city, threatened nature. 'Ever since the great planes were murdered at the end of the gardens', she felt that the city and its dark houses were 'watching the trees from dark windows', 'brooding' and 'simply biding their time' before felling them too.

She developed this theme in her poem 'The Trees Are Down', published in 1923, which is headed by a quotation from Revelation 7:3: 'and he cried with a loud voice: Hurt not the earth, neither the sea, nor the trees.' This was motivated by the felling of mature plane trees in Endsleigh Gardens, just to the south of Euston Station, and near her Gordon Street house, ready for the building of the large neo-Georgian headquarters of the Quakers.[49] The opening verse emphasizes the brutal nature of the felling of old plane trees she had known for half her life:

They are cutting down the great plane-trees at the end of
    the gardens.
For days there has been the grate of the saw, the swish of
    the branches as they fall,
The crash of the trunks, the rustle of trodden leaves,
With the 'Whoops' and the 'Whoas,' the loud common
    talk, the loud common laughs of the men, above it all.

She juxtaposes the unthinking heartlessness of the woodmen for whom felling was an everyday job with her anguished response. Her 'heart has been struck with the hearts of the planes', and when the men 'have carted the whole of the whispering loveliness away/ Half the Spring, for me, will have gone with them.'[50] Charlotte sent her friend Florence Hardy a copy of the poem in manuscript in 1924 and Florence replied saying that Thomas Hardy 'would love that' poem. But Florence also understood some of the practical problems of large garden trees and, thinking of their home at Max Gate, Dorchester, with its surrounding trees, she feared that if Thomas ever read it he would never allow boughs to be lopped off 'the trees that hem us round & make some of our rooms so dark & depressing'.[51]

Mew's poem establishes a strong contrast between her aesthetic response to the butchered trees and the masculine tree fellers. Thomas Hardy in *The Woodlanders* (1887), set in Blackmore Vale, Dorset, presents a landscape where the trees are not principally beautiful creatures, but ones that provide the productive economic fabric which sustains local society. Here both women and men work closely and directly with woodland products, but their physical work is in different spheres. When Marty South is first encountered she is seen in a cottage room 'seated on a willow chair, and busily occupied by the light of the fire, which was ample and of wood. With a bill-hook in one hand and a leather glove, much too large for her, on the other, she was making spars, such as are used by thatchers, with great rapidity.' On one side of the chair is 'a bundle of straight, smooth sticks called spar-gads – the raw material of her manufacture', on the other 'a heap of chips and ends – the refuse' which fuels the fire, and in front of her 'a pile of the finished articles'. Her task demands precise, dextrous skill and is set in a domestic scene.[52]

Marty also worked outside in the woods, especially assisting tree planting. She helped the woodman Giles Winterbourne to plant 'a thousand young fir-trees' in a 'spot which had been cleared by the wood-cutters'. She took a much less active role in this task than Giles: Marty 'performed the part of keeping the trees in a perpendicular position while he threw in the mould'. It was a cold day and Hardy notes, 'from the nature of their employment, in which he handled the spade and she merely held the tree, it followed that he got good exercise and she got none.' She had to ask permission, breaking the rhythm of planting, to 'run down the lane and back again' to warm her feet. But Giles's planting was also delicate and dextrous. His 'fingers were endowed with a gentle conjuror's touch in spreading the roots of each little tree, resulting in a sort of caress, under which the delicate fibres all laid themselves out in their proper directions for growth'. He made sure most roots were placed towards the southwest so that as the trees matured they would be better able to stand against 'some great gale blowing from that quarter'.[53]

The men carried out the heavier work of tree felling and hauling. These activities are dangerous and the trees can inflict serious and long-term injuries. The timber merchant George Melbury, as he gets older, can identify the reason for the 'stiffness about the arm, hip, and knee-joint' that became apparent as he walked about. They were caused by the 'sprains and over-exertions that had been required of him in handling trees and timber when a young man'. He remembered 'the origin of every one of these cramps: that in his left shoulder had come of carrying a pollard, unassisted, from Tutcombe Bottom home; that in one leg was caused by the crash of an elm against it when they were felling; that in another was from lifting a bole'. They are injuries caused by the active resistance of the trees being felled and carted. The trees also fight back psychologically. Marty's father, John South, a woodman aged 55, had become ill with worry over a tall shrouded elm which he could see from his bedroom window. When the wind blew 'the tree rocked, naturally enough . . . the sight of its motion and sound of its sighs had gradually bred the terrifying illusion in the woodman's mind that it would descend and kill him.' He had grown up with the tree and remembered as a 'little boy' thinking of 'chopping it off with my hook to make a clothes-line prop'. But he delayed and forgot and 'at last it got too big, and now 'tis my enemy' and will 'dash me into my grave'.[54]

The young poet Edward Thomas published at the age of nineteen in 1897 a collection of prose entitled *The Woodland Life*. In 'A Pinewood near London' he laments the felling of elms and the rapidity of their demise: 'Great rugged elms stretch side by side on the short turf of the meadow which yesterday they shadowed. The labour of a few short days, with many strokes of axe and saw, has felled them all.' They were felled just as they were about to spring into life: 'as the innumerable buds were beginning to show a faint purple when the sun gleamed among their branches'. The only trace that will remain when they have been removed will be 'great rounded grooves across the meadow' in the 'yielding soil, not yet hardened by frosts'.[55] In his late poem 'Bob's Lane' (1916), Farmer Hayward is 'an ambiguous patriarch, at once an earth spirit, liking women and loving horses'. 'Also he loved a tree./ For the life in them he loved most living things,/ But a tree chiefly. All along the lane/ He planted elms.' Many years later, long after Bob died, no one travelled along the lane because 'the mist and the rain/ Out of the elms have turned the lane to slough/ And gloom, the name alone survives, Bob's lane.' The critic Stan Smith argues that this 'is an omen . . . of an England changing for the worse, as a result of acts which are intended to be beneficial but actually destroy'. Moreover, when the poem was written in June 1916 'the slough and gloom of a lane which was once named and claimed by no one, could not but call up the desolation of another No Man's Land, in Flanders.'[56]

Forestry women

There is relatively little documentary evidence of women working in woods or with trees in the medieval period, although there is no doubt that they did so. An analysis of medieval coroners' inquests and manorial court records indicates that men were 'much more likely than women to die in fields, forests, mills, construction sites, and marl pits'. There were some dangerous activities and even collecting fire-wood, a vital task often carried out by women and children, had its perils. One inquest reported that a woman 'went out early in the morning to get kindling and climbed onto a tree leaning over the common way and fell'. Legal records hint at certain activities. In 1294 Mabel the merchant was charged at Chalgrave, Bedfordshire, with 'taking ash trees'.[57] There is more evidence of the types of tree and woodland work carried out by women in the seventeenth century. They were hired as 'labourers' for 'menial tasks' such as weeding in orchards and gardens. Accounts for Cardinal Wolsey's Hampton Court Palace dated 16 July 1516 provide a list of weeders in the orchards and most are women.[58]

Women landowners are sometimes identified as having an enthusiasm for trees and their management. One such is the 'great planter' Lady Lindsey, who was much involved in the design and layout in the 1670s of her son Robert Bertie's estate at Grimsthorpe Castle, Lincolnshire. She established a 'plantation of four square blocks of forest trees, and a further broad avenue extending into the far distance between a park and hayfield, each seen through a single row of forest trees'.[59] Her son later employed the garden designer Stephen Switzer to modify his mother's woodlands and avenues. Switzer recorded that the countess 'was reputed to be a continual Attendant and Supervisor of her Works, without any regard to the rigid Inclemency of the Winter-season'. He was fascinated to report that 'on the Measuring and Laying out the Distances of her Rows of Trees, she was actually employed with the Rule, Line, etc.' Her plantations were 'considerable Monuments of her Care and Pains', providing 'Umbrage and Safety' and 'great Advantages' to 'immemorial Posterity'.[60]

The indomitable Barbara Savile (1660–1734), the daughter of a wealthy merchant from Newcastle upon Tyne, provides a rare, documented example of an elite woman managing woodland. She married John Savile, a clergyman, who died in 1701, and their son inherited the Savile estates in Nottinghamshire and Yorkshire and later the baronetcy. Barbara Savile helped her son manage these estates for more than thirty years. She had a deep involvement in woodland management, including the identification of buyers and the negotiation of contracts for timber and charcoal, on the large, 17,000-acre Rufford estate, Nottinghamshire, much of which lay within Sherwood Forest. She was a tough negotiator, helping her son Sir George in finalizing contracts for wood sales. The importance of wood to the economy and cultural identity of the Rufford estate cannot be overestimated. Rufford's timber reserves were a capital resource worth over £21,000 in 1720 and annual coppice wood fellings, together with timber sales, brought in an average income of £532 during Sir George Savile's ownership, equivalent to just over a third of the estate's rental value. We can trace the powerful influence of Barbara Savile in 'The Sale of Woods, and other bargaining' through surviving family and estate correspondence.[61]

At Rufford two strategies were used for the sale of timber. Either the wood was sold after felling directly to the trades that required them, or it was sold as standing wood to middlemen, who then worked up the timber and sold it on. A letter sent from Rufford in March 1711 concerning a large intended fall of timber in the estate's New Park captures Barbara Savile's practical and assured manner of engagement with her son's woodland business affairs and their relationship to wider estate management concerns.[62] She informed and cautioned Sir George, who was in London, about a potential new partnership of wood chapmen she had learned about, suggested a strategic approach to the sale, and offered advice as to how the felling might proceed without damaging the aesthetics of the estate.

She told Sir George that speed was essential as Mr Ball, who was employed to value the Rufford timber, 'adviseth that you sell some part at Least by cutting it first down & so sell each tree in your new park to different trades, as the bark to taners, the Cordwood to forge men &c'. She also reminded Sir George of the impact that the timber fellings would make on the landscape of the parkland and the estate, offering felling advice that would minimize the visual disruption: 'I think you said you would look over the south Wing of the new park again ere they were took down haveing an aprehention they wud be Left too thin: from which reason, & because twill be most out of sight, Why may not the Trees far back in that park be first sold pray say'.[63]

In a letter of 1718, the Rufford steward Thomas Smith referred to an occasion when 'Mr Hayford and Mr Watts concluded ye Bargain [for cordwood] at Xmas [26 December 1715] in ye Dining room at Rufford at about Two o'clock in ye morning' and the subsequent 'caution' both madam and master issued to him not 'to take Down trees that have bark on before ye season of peeling'.[64] Denis Hayford (1635–1733) was a nationally significant iron and steel master. He had ironworks at Rockley, Rotherham and Cannock, and by 1680 supplied rod and bar iron to the Lancashire, Sheffield and Birmingham markets. In 1698 he acquired the Sitwell works at Staveley and Renishaw in Derbyshire.[65] He was a tough negotiator and in 1710 was involved in a dispute with the Duke of Newcastle over the supply of birch cordwood from Birklands in Sherwood Forest and Welbeck Park. The vivid description of the early morning discussion at Christmas shows Barbara Savile directly involved with her son in negotiating with Hayford about the necessity that they should be able to peel and harvest the valuable tanning bark before Hayford felled the trees and took away the timber.

The 'Christmas' contract bound Sir George between 1715 and 1719 to supply a specific volume of charcoal-grade wood to an iron forge at Carburton. By November 1718 the supply was in serious arrears and no obvious source of cordwood was identifiable on the Rufford estate. For reasons unspecified in the correspondence, Sir George had both failed to instruct his steward over the measures to be taken and was unavailable. Under these difficult circumstances, the Rufford steward Smith turned to Barbara Savile. She had been present at the December 1715 meeting and had clearly been in communication with Smith already over the cordwood assignment and taken measures to ascertain the supply margins. Smith sent an urgent request for advice to Barbara Savile at her London house in Golden Square on 26 November: 'I Do not know where youll Get what he Expects, thers none y^t I Can hear on to be bought nor would y^t please because our buying only prevents them buying it. Besides ye article says y^t its to be found on some part of my masters Estate.'[66]

Five years after the Hayford deal, Barbara Savile was at Rufford finalizing the terms of another significant wood sale with Mr Cotton, an ironmaster and wood dealer, in the absence of her son. Correspondence between Sir George and his Yorkshire steward, William Elmsall, makes clear that this was a controversial deal, the terms of which were negotiated over several months: Elmsall strongly favoured a closed auction; Savile accepted Cotton's initial bid without inviting competitors such as Hayford. Notably, it was Barbara Savile who closed the deal. In May 1721 Cotton arrived at Rufford with the contract and met Barbara Savile in the presence of the Rufford steward, Thomas Smith, who reported the transaction to his absent master: 'Mr Cotton is very willing to Comply to any thing that may make you & Madam Savile easy ... he is so very fare in every respect that Madam Savile & my Self are perfectly Satisfied all will be as Safe & Right as is possible without ye Least appearance of Danger or hazard.' It is clear that Barbara Savile was ready to take decisive action as her son's proxy.[67]

It has recently been argued that until the First World War in Britain 'it was virtually impossible for a woman of any class to have what may be described as a career in forestry.'[68] The membership list of the English Arboricultural Society for 1890 records no female members.[69] The Swedish forest historian Lars Östlund notes that 'Logging and forestry have traditionally been seen as a purely masculine sphere.' The 'timber-frontier, which swept across northern Europe and North America' from the late nineteenth century onwards demanded many male workers. There was also some work for women 'in the Nordic countries during this time; they worked in silviculture, they maintained farms while men worked in the forest, and they were cooks in forest camps.' The work of female forest camp cooks was 'indispensable for the new forestry system which was built up around collective house-holds, and it enabled the male forest workers to focus on other tasks.'[70]

It might be expected that some women would be employed in lighter woodland work such as harvesting osiers grown in willow beds. But a 1926 study carried out for the Agricultural Economics Research Institute, Oxford, by Helen Fitzrandolph and Mavis Doriel Hay found that their work was limited to the preparation of the osiers: 'Men are employed for the cultivation of the beds and the cutting of the rods, women for peeling both white and buffed rods, and sometimes for additional help with weeding the beds.' Buffing involved the boiling of 'rods for from two to five hours'. The rods were left to cool overnight. They were 'peeled the next day by women, who stand around the tank picking out the still hot rods as they need them. The peelers wrap rags around their fingers to prevent soreness, and remove the bark by pull-ing the rods through their hands.' There were extensive osier beds in the Trent Valley and here 'women, helped by boys and girls, are em-ployed on peeling buffed rods throughout the winter until April.' Although the work was 'hot and dirty', it was 'said to be healthy, the boiled rods giving out a pleasant acrid smell of tannin'. However, like 'all women workers in agriculture', the 'social status' of female willow workers was 'generally considered as below that of domestic servants'. The academics Fitzrandolph and Hay felt that such work 'naturally tends to attract the hardier type of woman because of the exposure to weather and the hard wear on clothes'.[71]

In the Second World War women were employed in a wide range of woodland work. By the end of 1941, 3,800 members of the Women's

Land Army were working in woods and forests and in April 1942 the Women's Timber Corps was established. The following year 8,500 women were working for the Forestry Commission and the timber trades. Most worked at practical woodland management felling trees, sometimes over 80 feet (24 m) high, using axes weighing 4½–7 lb (2–3 kg) and two-handed saws. Mavis Williams provides a lively account of her life as a 'Lumber Jill' working for the Women's Timber Corps from 1942. A six-week training course was provided by the Ministry of Supply on the edge of the King's Forest at Culford, Suffolk (illus. 34). They were trained in driving, haulage, loading and unloading timber, and 'calculating, recording, assessing volume, wages,

34 Land Girls using a double saw to cut down a tree as part of their training at the Women's Land Army camp, Culford, Suffolk, 1943.

correspondence, communications'. They also had practical training in 'felling, lopping, stripping, sawing, stacking, loading, charcoal making'. Once trained, she worked at various sites in Herefordshire, Shropshire, Cornwall and Devon. The women 'worked besides men – felling, sawing, stacking, cutting logs by hand or on the Porta-saws, packing kilns, and hauling wood to the sawing ramps'.[72]

In 1944 Mavis was working at Dunmere Woods, near Bodmin, Cornwall. The women she worked with were aged between 18 and 22 and came from all around the country: Liverpool, Halifax, London and Bristol. Most had very little experience in forestry and their former jobs included a 'buyer for a fashion house' and sales assistants at lingerie and cosmetic counters. The work at Dunmere involved 'Cross cutting and sawing wood into various lengths for pit-props for the Welsh minefields'. They had to 'measure out square chains for the fellers fighting our way through the remaining undergrowth, crossing numerous streams, marking out the boundaries by felling a tree'. Payment was not too bad: 'The girls sawing pit-props and paid by the inch were now earning a lot of money as were the fellers and haulers.' There were four standard lengths of pit-props and 'the classification to different sizes' was the responsibility of the girls. The cut props are 'stacked on a slope above the mineral railway . . . the smaller props are heaved into' the empty trucks. Larger props were rolled down skids into the trucks. The *Cornish Times* reported, 'These girls take a pride in their work, and they are doing a most important war job, and even when peace comes wood and timber of all sorts will be in urgent need for all our bombed cities and for industries.' It noted that the 'Timber Corps will be busier than ever then, and the girls perhaps busier still'. However, after the war 'replanting went on apace, but this was often done by men. Dunmere in Cornwall was replanted immediately after the war and is now, after fifty years, restored to its former beauty, full of bluebells and primroses in Spring.'[73]

Although most women stopped doing practical work in the woods after the war, some were employed to work for a Forestry

Commission census of private woodland: 'Several groups spread out across each county armed with six-inch maps, searching out all woodland over five acres.' They recorded 'species, age and condition, of the remaining trees, the size of areas which had been felled and the acreage of each and every species'. This task took two and a half years to complete.[74] The Forestry Commission reported in 2023 that 'Forestry work has historically attracted fewer female candidates than male candidates,' this being 'particularly the case in forestry operational roles'. However, by 2022, 44 per cent of the staff employed by the Forestry Commission were female and a growing number are engaged in practical forestry activities including tree-felling.[75]

6

ARBOREAL AESTHETICS

Trees and forests have for centuries, probably always, been valued by humans for their aesthetic qualities. For many authors their inherent beauty is so obvious that they feel no need to discuss it. The early twentieth-century Irish tree enthusiast Augustine Henry, when writing on the relationship between trees and health just after the First World War, pointed out that 'A great deal might be said in praise of the aesthetic value of forests and trees, which influence favourably the spirits and consequently the health of the people; but a mere mention of this aspect of the subject is sufficient. Few will deny its importance.' He thought that one of the 'best arguments for locating sanatoria for tuberculous patients in forest districts' was that they provided excellent wind breaks, allowing patients to walk in the hospital grounds. Moreover, the greatest 'hygienic asset of the forest is the purity of the air', as 'Smoke, particles of dust, injurious gases, and bacteria which are all found in the air of cities, are rare or completely absent in that of forests.' Nineteenth-century designers of sanatoria stressed the need for a careful balance between open land and trees. Treeless sites were at a great disadvantage, but 'the trees must not be crowded around the building so as to interfere with ventilation.' Dr Kincaid Etlinger of the Pinewood Sanatorium, Wokingham, which opened in 1901 as the London Open Air Sanatorium, noted that when he arrived the 80-acre site was 'densely covered with pines, growing close up to the buildings on all sides'. The atmosphere was 'gloomy and depressing', so he 'heroically cleared' the sanatorium grounds 'to as great a distance as possible', allowing the sun and wind to improve the atmosphere.[1]

The value of forests as recreational and 'therapeutic landscapes' became increasingly important through the twentieth century. Cycling though woodland scenery was popular before the First World War: E. M. Forster noted on 28 May 1905 that the 'woods are full of bicyclists' paths' in the mountains around Daber (now Dobra, Poland).[2] Officers convalescing at Craiglockhart military hospital, Edinburgh, including the poets Siegfried Sassoon and Wilfred Owen, were encouraged to take walks in the nearby woods. Camping and caravan sites within woodland became very popular in the twentieth century, with thousands of sites being established such as the Bracelands camping site in the Forest of Dean, which was established in the 1950s, and many sites in the Landes, southwest of Bordeaux, as in the coastal pinewoods at Vieux-Boucau-les-Bains (illus. 35).

35 Campeurs, Vieux-Boucau-les-Bains, Landes, c. 1959, postcard.

Commercial recreational and holiday facilities such as Center Parcs, golf courses and hotel complexes are increasingly built within forests. It sometimes feels that there is scarcely a wood in Britain that does not have a glamping site, sculpture walk or orienteering map. Carefully categorized mountain biking trails, ski runs and reinvented pilgrim routes such as the Camino de Santiago and Via Francigena encourage travel through woods and forests, although this may compete and often conflict with some traditional woodland activities such as hunting and shooting.

The health and well-being benefits of trees, woods and forests are increasingly central to forest policy and practice. Approaches such as 'green care', 'wilderness therapy' and 'green exercise' are frequently recommended.[3] Doctors relatively routinely 'prescribe' a 'dose of nature' to support mental health and alleviate anxiety, fatigue and depression encouraged by 'modern' lifestyles. Such doses entail individual exposure to, or immersion within, 'natural' environments such as woodlands. They draw on the 'restorative therapeutic' values of these environments through visual aesthetics and processes that encourage practical encounters with the 'natural' world.[4]

The appearance and scents of tree bark can fascinate and almost infatuate. The inveterate gardener E. A. Bowles noted in his diary on 1 January 1902 that he 'planted narcissus bulbs in the grass bordering the drive' and 'Scraped Yew till 5 p.m.' There was a row of yews along the drive and he 'greatly admired the stout pillars of their boles'. Every year he stripped

> them of the loose bark to reveal the shining mahogany
> within. Having scraped, he then scrubbed, fetching
> a chair and standing upon it to make them gleam higher.
> The final touch was a bucket of water thrown at each,
> when he would stand back to admire them, rich with
> colour in the sunshine.

The bark of the grand fir (*Abies grandis*) has little blisters that eject a very sticky sap when squashed. This has an attractive, strong, grapefruit-like scent, which a woodman told me his girlfriend preferred to his normal aftershave.[5]

A recent study, '"Hey, tree. You are my friend"', examined contemporary values associated with nature by inviting 'visitors to a natural area to write letters to non-human elements of the ecosystem (trees)'. This was carried out at the Intervale Center, which consists of 360 acres of 'farmland, trails and open space along the Winooski River' in Burlington, Vermont. The centre was established in 1988 to 'enhance farm viability and land sustainability'. They received by email 45 letters from 25 different writers. There were many straightforward questions about the site and the different tree species found there. Some people focused on a tree's appearance: 'There's nothing like standing right in front of a Hackberry tree to really see the bark texture.' Other writers appeared to take a more Whitmanesque stance: 'I want to let you know how stoked I am to get a chance to talk with you. I've never had a tree friend before.' Or 'I relate to your love of life and endurance. You have reached for the sun for many years which is what I'm doing too.'[6]

Walt Whitman at Timber Creek

There could hardly be a keener advocate of the therapeutic value of trees and woods than the American poet Walt Whitman (1819–1892) (illus. 36). He was brought up on Long Island, where he relished the shorelines and views of the ocean. He preferred these to the 'extended wide central tracts of pine and scrub-oak' where charcoal was made, which he found to be 'monotonous and sterile', although he enjoyed 'inhaling the peculiar and wild aroma' on his lengthy cross-country walks.[7] He worked as a journalist mainly in New York through the 1840s and '50s and published the first of many editions of *Leaves of Grass* in 1855. He suffered a first stroke in 1858 and moved to

36 George Collins Cox, *Walt Whitman*, 1887, photograph.

Washington in 1863. During the Civil War (1861–5), and afterwards, he regularly visited hospitals as a volunteer, nursing injured and ill soldiers, making many notes describing the war's 'hell-scenes' and the bravery of the soldiers. In January 1873 he had another stroke and started to share a house in Camden, New Jersey, to the southeast of Philadelphia, with his brother George. He also went 'for weeks at a time, even for months, down in the country to a charmingly recluse and rural spot along Timber creek, twelve or thirteen miles from where it enters the Delaware river'. He called himself a '*half-Paralytic*' and it was 'to my life here that I, perhaps, owe partial recovery'.[8]

While at Timber Creek he continued to write and make notes and jottings of his ideas and thoughts, some of which he published in 1875, with a full American edition in 1882, followed by a British one in 1887. These carefully orchestrated *Specimen Days* give the false impression of being a diary, frequently providing a date, or a time, but are often vague about the year. They document the value of trees and woodlands in his recovery from his wartime experiences and strokes. He found 'the woods in mid-May and early June my best places for composition. Seated on logs or stumps there, or resting on rails, nearly all the following memoranda have been jotted down.' The woods provided 'Dear, soothing, healthy, restoration hours – after three confining years of paralysis – after the long strain of the war, and its wounds and death'. He identified the beauty of particular trees in detail: 'As I write this . . . I am sitting near the brook under a tulip tree [*Liriodendron tulipifera*], 70 feet high, thick with the fresh verdure of its young maturity – a beautiful object – every branch, every leaf perfect.' When the tulip tree was in flower 'it swarms with myriads of these wild bees', who sought 'the sweet juice in the blossoms' and 'whose loud and steady humming make an undertone to the whole'.[9]

On 1 September 1876 Whitman summarized the lesson that humans could learn from a tree. He decided not to choose 'the biggest or the most picturesque tree to illustrate it', but instead selected one of his favourites, 'a fine yellow poplar, quite straight, perhaps 90 feet high, and four thick at the butt. How strong, vital, enduring! how dumbly eloquent! What suggestions of imperturbability and *being*, as against the human trait of mere *seeming*. It *is*, yet says nothing.' Although trees did not speak or write they 'do a great deal better'. Whitman tells us to 'Go and sit in a grove or woods, with one or more of those voiceless companions, and read the foregoing, and think.' On 2 June the following year he speculated what would happen if trees could speak and walk. He noted, while sitting under 'my great oak', the nearby tulip tree – 'the Apollo of the woods – tall and graceful, yet robust and sinewy, inimitable in hang of foliage and throwing-out of limb; as if the beauteous, vital, leafy creature could walk, if it only would'. He had a 'dream-trance' that his 'favourite trees step out and promenade up, down and around, very curiously – with a whisper from one, leaning down as he pass'd me, *We do all this on the present occasion, exceptionally, just for you.*'[10]

In some ways Whitman's tree aesthetic is centrally picturesque and his descriptions could have been made by William Gilpin or Uvedale Price. Noting the light on 4 August at 6 p.m. he sees the 'rare effects on tree foliage' of the evening sunlight:

> The clear beams are now thrown in many new places, on
> the quilted, seam'd, bronze-drab, lower tree trunks, shad-
> ow'd except at this hour – now flooding their young and
> old columnar ruggedness with strong light, unfolding to
> my sense new amazing features of silent shaggy charm,
> the solid bark, the expression of harmless impassiveness,
> with many a bulge and gnarl unreck'd before.

He is so struck by the effect that he recalls fables of 'people falling into love-sickness with trees'. But when 'loafing in the woods' in March and 'seated on a log in the woods, warm, sunny, midday', the woodland takes on a more austere, sublime character after a walk 'deep among the trees, shafts of tall pines, oak, hickory, with a thick undergrowth of laurels and grapevines – the ground cover'd everywhere by debris, dead leaves, breakage, moss – everything solitary, ancient, grim'.[11]

In addition to the emotional and psychological benefits of being among trees and in woodland, Whitman emphasized their practical value as an outdoor gymnasium. On 5 September at 11 a.m. he wrote a note 'shelter'd under a dense oak by the bank, where I have taken refuge from a sudden rain'. He had been at his 'daily and simple exercise I am fond of – to pull on that young hickory sapling out there – to sway and yield to its tough-limber upright stem – haply to get into my old sinews some of its elastic fibre and clear sap'. In addition he selected

> strong and limber boughs of beech and holly, in easy
> reaching distance, for my natural gymnasia, for arms,
> chest, trunk muscles . . . I . . . wrestle with their innocent

stalwartness – and *know* the virtue thereof passes from
them into me. (Or may-be we interchange – may-be
the trees are more aware of it all than I ever thought.)

His fascination with and love of trees also made him ponder the long-
term implications of tree loss and he stressed that 'the tree question
will soon become a grave one.' This was especially true for the prairies:
'stretches of hundreds and thousands of miles where either not a tree
grows, or often useless destruction has prevail'd'. He concluded that
'the matter of the cultivation and spread of forests may well be press'd
upon thinkers who look to coming generations of the prairie States.'[12]

Whitman stayed as a paying guest with the Staffords, a farming
family, at Laurel Springs, New Jersey. Gradually, as his condition
improved, 'he became more mobile and independent, and he would
spend many hours' at the creek 'resting, observing, musing, jotting,
and practicing a program of physical therapy that partially rehabilitated
his body'. The physical therapy included 'sun-bathing, mud-bathing,
bathing at a flowing spring, scrubbing his skin with a hard brush, saun-
tering along the bank wearing only shoes and a straw hat, singing bits
of opera and folksongs and reciting poetry, and wrestling with the
saplings that grew along the bank'. Most biographers have emphasized
Whitman's solitude at Timber Creek, but he also had many friends
there including Mrs Stafford, who 'had a special fondness for Whitman',
and her son Harry, who 'became one of the most intense attachments
of his life'. In the mid-twentieth century the site where Whitman
bathed and pondered 'lay beneath the town dump, but through the
efforts of local citizens it was cleaned up and restored' and is now a
small park and tourist attraction named Crystal Springs Whitman
Walk. It is closely bounded by a grid of suburban houses on streets
named Poplar, Maple and Sycamore Avenue.[13]

E. M. Forster's Piney Copse

E. M. Forster (1879–1970) is not usually celebrated as a nature writer, although his early short stories and diaries show he was deeply sensitive to arboreal aesthetics. In 1947 he recalled the 'first story I ever wrote and the attendant circumstances remain with me vividly'. When on holiday in May 1902 he 'took a walk near Ravello' and 'sat down in a valley, a few miles above the town, and suddenly the first chapter of the story [*The Story of a Panic*] rushed into my mind as if it had waited for me there'. A group of English tourists climb above Ravello where 'the valley and the ravines and ribs of hill that divided the ravines were covered with leafy chestnut' looking like 'a many-fingered green hand, palm upwards, which was clutching convulsively to keep us in its grasp'. The tourists bicker about the scenery, and the artist Leyland is horrified at some felled trees, to him 'the mere thought that a tree is convertible into cash is disgusting . . . the woods no longer giving shelter to Pan.' The narrator felt the 'cutting was very necessary for the health of the larger trees.' The panic of the story concerns the young boy Eustace, who becomes uncontrollable and when back at the hotel 'alighted in an olive tree, looking like a great white moth, and from the tree he slid on to the earth . . . uttered a strange loud cry . . . and disappeared among the trees'. In 1903 'a hollow tree not far from Olympia' inspired Forster to write 'The Road from Colonus'. The 'huge' plane trees 'magnificently clothed in quivering green' instigate Mr Lucas's desire to connect with Greek countryside and culture.[14] On 8 April 1904, while staying with his uncle at Felton, Northumberland, Forster noted. 'In the woods this afternoon I felt how unwarrantable it is to conclude that trees and flowers have no feeling and are insensitive to procreation and gestation.' He thought that 'Breaking a twig in spring is like cutting into flesh: I ate the bud of a horse chestnut, and it was overpowering, like animal tissue.'[15]

The following year Forster stayed with Elizabeth von Arnim, author of the bestselling novel *Elizabeth and Her German Garden* (1898), as a tutor to her children at Nassenheide, Pomerania (now Rzędziny, Poland). On one trip in May 1905 they went 'through spindly oaks to Falkendorf where I saw two most beautiful things: bathers running naked under the sun pierced foliage and a most enormous beech, standing in the village like a god. A villager was proud of it.' These notes hint at a key scene in Part Two of his novel *A Room with a View* (1908) set in Summer Street (based on Holmbury St Mary), Surrey. Windy Corner, the home of the Honeychurch family, had fine views of 'pine-clad promontories descending one beyond another into the Weald'. The characters Freddy Honeychurch, George Emerson and Mr Beebe bathe in 'The Sacred Lake', a small woodland pond with 'a slippery bank of pine needles' among 'the pine-trees, rising up steeply on all sides, and gesturing to each other against the blue. How glorious it was!' This part of Surrey was becoming popular with wealthy Londoners who built houses among the pines, which partially protected them from 'poverty and vulgarity, for ever trying to enter, just as the London fog tries to enter the pine-woods, pouring through the gaps in the northern hills' (illus. 42).[16]

Not long after the First World War Forster bought a small 4.5-acre wood named Piney Copse in Surrey. His writings about this wood throw light on the complex motives and interests brought about by woodland ownership. His purchase is strangely foreshadowed in his story 'Other Kingdom' of 1909, in which Harcourt Worters buys a beech wood, 'Other Kingdom Copse', to try, unsuccessfully, to capture his fiancée.[17] Forster's wood lay just to the west of his mother's house in Abinger in densely wooded Surrey. He recollected his purchase in his memoir *West Hackhurst: A Surrey Ramble*, written between 1943 and about 1960, which remained unpublished in his lifetime.

He had known the wood for many years but did not think much about it except that a path through it was a useful shortcut to Gomshall railway station. The wood had been 'planted with larches and scotch fir about 80 years ago . . . and I have talked to men who worked in it as boys'. The wood 'had always been tight and dull, like a German wood, and no one wandered in it, except for business'. During the First World War the best timber was felled. Consequently the wood now 'became dull and open, and a few straggling oaks, which had never expected to face the light of day, stood about in it uncertainly, and raised unhealthy antlers into the sky'. However, in the early 1920s the pressure to build new houses in Surrey was great, and Forster noticed 'two men digging a hole' in the wood. The rumour soon spread that a local farmer 'was going to build bungalows on it'. Forster was horrified as this development would spoil his mother's house. Consequently 'after the usual frights and crossnesses, I became for the first time in my life the owner of freehold property.'[18]

It was the financial success, especially in America, of *A Passage to India*, published in 1924, that enabled Forster to spend £450 on the purchase of the wood. In 1926 he wrote a drolly entertaining satire, 'My Wood, or the Effects of Property upon Character', for *New Leader*, the newspaper of the Independent Labour Party.[19] He stresses that 'It is not a large wood – it contains scarcely any trees, and it is intersected, blast it, by a public footpath.' He uses his purchase to examine 'the effect of property upon' the owner's character, suggesting three ways. 'In the first place, it makes me feel heavy. Property does have that effect.'[20] The second effect is that 'it makes me feel it ought to be larger.' The smallness of the wood, and the public footpath crossing it, made him feel threatened:

> The other day I heard a twig snap in it. I was annoyed at first, for I thought that someone was blackberrying, and depreciating the value of the undergrowth. On coming nearer, I saw it was not a man who had trodden on the twig and snapped it, but a bird, and I felt pleased. My bird.[21]

Third, ownership of the wood made him 'feel that he ought to do something with it. Yet he isn't sure what.' Sometimes he thinks he will cut down the few remaining trees, 'at other times I want to fill up the gaps between them with new trees.' But he thinks, 'Both impulses are pretentious and empty. They are not honest movements towards money-making or beauty. They spring from a foolish desire to express myself and from an inability to enjoy what I have got.' At the end of his article, he returns to the effect of trespass on the woodland and its owner. In addition to blackberrying, people pulled up foxgloves and 'ladies of an educational tendency even grub up toadstools to show them on the Monday in class. Other ladies, less educated, roll down the bracken in the arms of their gentlemen friends. There is paper, there are tins. Pray, does my wood belong to me or doesn't it?' He jokes he will fence it off and eventually 'Enormously stout, endlessly avaricious, pseudo-creative, intensely selfish, I shall weave upon my forehead the quadruple crown of possession until those nasty bolshies come and take it off again and thrust me aside into the outer darkness.'[22]

Forster wrote 'My Wood' almost immediately after his purchase, but his later writings show that he came to relish ownership: 'I didn't know at first what pleasure Piney Copse was going to be.' A neighbouring landowner 'gave me some saplings for planting – birch, mountain ash, a few beech, I stuck them in – they seemed no larger than buttercups and they all had to be wired against rabbits.' He began to become aware of the difficulties involved in establishing young trees. He then 'got some other stuff – sweet chestnuts, more beeches, a few horse chestnuts, a few wild cherries, crabs and conifers and stuck them about too'. Some failed and some grew better than he expected and 'have grown into young men, wild & charming, who touch one another unexpectedly, and give the place a romantic beauty it can never have had before'. The trees also provided heating fuel, indeed the 'drawing room fire, by which I now write, has been fed by them for years'. The wood also gave him some solace. When he stepped 'into its green shade, I annihilate the worries of the house, the garden, the war, and when the planes swoop as usual out of the infernal blue they threaten a more dignified death'. And he felt that being a freeholder in some 'inexplicable' way allowed him to achieve 'this particular poise'. He emphasizes 'the happiness, joy and strength I have found through purchasing Piney Copse. And the pride, for in replanting it I have helped to maintain England. But it hasn't made me more tolerant or more civilised.'[23]

The continuing importance of trees to Forster's understanding of landscape history is brought out in his script for the Abinger Pageant of July 1934, which was held to raise funds for preservation of the church. Tom Harrison, the producer of the pageant, had the idea of a woodland theme and Forster was 'much taken with this notion of a "Pageant of trees".'[24] Music for the pageant was composed by Ralph Vaughan Williams, who lived nearby. The script notes that 'Abinger is a country parish, still largely covered with woodland.' The pageant was held in the Old Rectory Garden, in which was a great tulip tree that 'according to tradition was planted there three hundred years ago by the diarist, John Evelyn.'[25] The narrator of the pageant is the 'Woodman', who welcomes visitors first 'to our woods, because they are oldest. Before there were men in Abinger, there were trees.' The pageant is about 'the history of a village lost in the woods. Do not expect great deeds and grand people here. Lords and ladies, warriors and priests will pass, but this is not their home, they will pass like the leaves in autumn, but the trees remain.' There are various historical scenes interspersed, with the Woodman reminding his audience that throughout the history of the parish 'the trees are growing, the trees are cut down.' The introduction of the larch, the silver fir and the Scots pine are celebrated and the Woodman questions, 'Which is better – that ancient Royal wood of Saxon and Norman, where the oak and the ash were king? Or this later wood of ours republican, where many trees mingle?' His answer is that 'I cannot tell you. I am only the Woodman, but I know that though the trees alter the wood remains.'[26]

The whole pageant is imbued with an enthusiasm for trees, their beauty, their historical resonances and their utility 'for charcoal, for the iron works, for the forge.'[27] But Forster's experience with the potential threat of building at his own Piney Copse and his concern over the destruction of countryside comes through strongly in his epilogue to the pageant. The Woodman tells the audience:

> Houses, houses, houses! You came from them and you must go back to them. Houses and bungalows, hotels, restaurants and flats, arterial roads, by-passes, petrol pumps, and pylons – are these going to be England? Are these man's final triumph? Or is there another England, green and eternal, which will outlast them?

Forster argues,

> If you want to ruin our Surrey fields and woodlands it is easy to do, very easy, and if you want to save them they can be saved. Look into your hearts and look into the past, and remember that all this beauty is a gift which you can never replace, which no money can buy, which no cleverness can refashion.[28]

In 1937 he contributed a chapter called 'Havoc' to Clough Williams-Ellis's polemical *Britain and the Beast* in which he laments that over the last fifteen years the countryside has been 'gashed to pieces with arterial roads'. As the population increases, 'Something has to decrease, and it has to be the woods and downs, the hedges and birds.' He thought civilization would be 'grateful to us for bequeathing a few samples of the countryside, of the beauty that took three hundred years to grow, and can never be replaced'.[29] Forster 'felt proud of his wood: even secretly patriotic, as though by means of it he were helping to maintain England'; when replanting trees, however, he avoided oaks 'because of their too-patriotic associations'.[30]

Forster continued to go to his wood after he had to leave West Hackhurst and Abinger in 1946 following his mother's death the previous year. In October 1950 he visited during a period of illness and 'lay in the wood and climbed and lay on downs which I may not be able to do again, with heart weak and left foot and leg swollen'. He records that he did not call on any of his former neighbours and friends – 'what a relief!' – but instead was delighted to visit the 'fine trees I've planted' in 'the square place in the upper part of the wood'. The 'sweet chestnut to east, laurels to south, elsewhere beech, birch, crab-apple, whitebeam, horse chestnut' had become more real to him than his erstwhile Abinger acquaintances. When he bequeathed his wood to the National Trust in 1970, 'together with a gift of £100', he was able to ensure at least this small plot was secured from development and from the creeping 'red rust' of London that his character Margaret Schlegel had seen in *Howard's End* (1910) 'in Surrey and even Hampshire now', and who worried that 'London is only part of something else, I'm afraid. Life's going to be melted down, all over the world.'[31]

Therapeutic Riviera palms

The widespread presence of palm trees gives stretches of the Ligurian coast an almost tropical character, especially around Bordighera and Sanremo. These coastal towns became major health resorts in the late nineteenth century and international tourists and residents celebrated this exotic landscape. Palms proliferated in gardens, as street trees and in horticultural collections. Two species of palm are native to Europe, *Phoenix theophrasti*, restricted to Crete, and the more widespread *Chamaerops humilis*, found in parts of Spain, France, Italy and North Africa. The range of the latter was reduced in the nineteenth century when it became a target for plant collectors and amateur botanists. The date palm *Phoenix dactylifera* has been cultivated in Asia and Africa for several thousand years and transported around the Mediterranean by the Greeks and Romans.[32] Palms have become signature trees for wealthy tourist landscapes in the Mediterranean and around the world from the Caribbean to the Maldives. On the Italian Riviera they are frequently found in the grounds of villas and hotels, at railway stations and in public squares.

Palms, particularly *Phoenix dactylifera*, were once an important element of the local economy, cultivated especially for palm fronds to be sold for religious festivals. They have been cultivated in Liguria since at least the fifteenth century. The lands of Sanremo were described in 1453 as 'rich in lemon trees and palms, the rarest trees in Italy' (*palmaeque, arborum in Italia rarissimarum*).[33] In the eighteenth century the Val di Sasso, just to the east of Bordighera, had dense groves of palms growing on terraces which were depicted by the Genoese cartographer Matteo Vinzoni. By the nineteenth century the valley specialized in the growing and making of *palmizi*, the braided palm branches blessed on Palm Sunday and distributed as a sign of devotion.[34] The production of *palmizi* was labour intensive. Farmers had 'to take extra care in tying the branches from the top in order to preserve the central shoot from the rays of the sun, removing the green colour and giving them that whiteness which is considered highly valuable'. The branches were tied up in June and July and untied in December, when the bundle of palm fronds at the top of the tree was cut off and placed in a dark room. They were 'sent to Rome by the end of Carnival'.[35]

An English resident, Frederick Hamilton, noted that 'turning the Palm leaves into a yellowish white colour, is produced by a process exactly similar to that employed by the market gardeners in obtaining the white centres of the lettuces.' The method was described in an interview by one of the last *parmura* of Bordighera, Franco Palmero, in 2015. The *parmura* climbed the palms with a rope made of hemp and cut the fronds growing outwards so he could reach the centre of the tree, where he could stand. The fronds were tied using a rope made of *Clematis vitalba*, which was collected from the woods and softened by soaking in water.[36] A painting of 1883 by the German painter Ascan Lutteroth (1842–1923) shows a group of palms, two of which have their fronds tied up (illus. 43).

The palm fronds used in the churches of Rome were called 'Roman style'. They were 'sorted in bundles of 60 shoots each, big and small', and the price of each bundle was between 20 and 30 lire. After the palm had been cropped the branches were 'left in their natural state for a year so that the tree increases in strength and regenerates new shoots'. There was a distinct market for Jewish ceremonies that required green palm fronds, and so the leaves were not tied up and blanched. At Sanremo, as tourism increased and gardens and villas spread, 'all the places where the palm trees were cultivated became ornamental and the plants were no longer subjected to binding.' In Bordighera palm cultivation remained profitable until the 1960s, after which many terraces were abandoned and palm trees remained uncut or were replaced with other crops such as asparagus, mimosa and eucalyptus for decorative foliage. Moreover, large old trees themselves became a valuable commodity and were dug up and sold for ornamental use.[37]

Palms were greatly valued by the increasing number of foreign visitors to the Riviera in the 1850s. Alfred Tennyson in his poem *The Daisy* (1853), written after travelling with his wife along the coast, celebrated the palms they had experienced:

> O love what hours were thine and mine
> In lands of palm and southern pine
> In lands of palm, of orange blossom,
> Of olive, aloe, and maize and vine

But he did not find the cultivated palms, 'the clipp'd palm of which they boast', pleasing. The popularity of the Riviera for British tourists was enhanced by the runaway success of Giovanni Ruffini's novel *Doctor Antonio*, first published in English in 1855. Ruffini (1807–1881) was born in Genoa and, being a keen supporter of the revolutionary republican Giuseppe Mazzini, lived in exile in London and Paris for much of his life. He had worked with Donizetti on the libretto for *Don Pasquale* in 1843. *Doctor Antonio* is set in 1840 and recounts a platonic love story between a revolutionary doctor and Lucy, the wealthy daughter of an English baronet. Lucy is fascinated by the palms: 'What an Eastern look those waving palms give the hill of Bordighera! One might believe one's self in Asia Minor.' Later, while visiting Sanremo, she asks, 'pointing to the plantations that covered the shore', whether 'palm-trees grow naturally in this part of the country? . . . or are they cultivated for beauty's sake?' Antonio replies that their beauty 'is their least merit in the eyes of their proprietors' and that palms 'are a very profitable kind of property, and that is why they are cultivated.'[38]

The publication of Henry Alford's influential guide to the Riviera in 1870 perhaps marks the point when the palms of Bordighera became themselves one of the reasons for visiting the Riviera. Alford, Dean of Canterbury, was a biblical scholar, poet and friend of Alfred Tennyson. In February 1869 he set off for 'six weeks alone with Nature' after a hard winter's labour in charge of 'the enthronement of our new Archbishop' and published his tour in 1870 as *The Riviera: Pen and Pencil Sketches from Cannes to Genoa*. Alford emphasized that 'The palm-glory of Bordighera is not to be seen without going up into the town, and beyond the town' where the 'noble' palms 'almost gird it round on the western and northern sides, and grow in profusion – in coppices and woods – of all sizes, from gnarled giants of 1100 years' reputed age, to little suckers which may be pulled up by hand, and carried to England'. He stressed the aesthetic aspects of the trees and thought that 'there is no end to the picturesque groupings of these lovely trees, and their graceful effects in the sunlight.'[39]

From an artist's perspective he felt that the palm was 'the best purveyor of flecked and dancing shade'. He used the vocabulary of the palette to identify the special advantages of the palm: 'the yellow, and the pale green, and the rich burnt sienna of the various foliage; the rough deep markings of the rich brown stem; and now and then the burning chrome of the fruit stalks hanging in profuse clusters'. These visual attractions were supplemented by the sounds of the leaves: 'the silvery whisper of reeded fronds which dwells everywhere about and under it'. Alford outlined a precise itinerary that visitors should take through the town to take in the best views of palms (illus. 44). One looked over 'some old rough battlements', which allowed a comparison between tree forms: 'several graceful palms, and a cypress in marked contrast, with its solid vertical dark spire, to the feathery palm-fronds gleaming against the heavenly blue'. For another view the artist had to walk through the arch and take a 'stand a little above a baker's shop' and 'in the middle of the paved gangway', which must have annoyed local residents. He found some ancient garden palms, not cultivated for their *palmizi*, 'loaded with their dazzling fruitage: we pick up and eat the soft sweet dates which strew the ground.' He also emphasized the biblical connotations of the palms. When he reached home in Canterbury and looked back on this 'idyll of palms' and 'cut the edges of his sketches', he dreamt of 'a certain procession . . . down the stony path from Olivet' and 'of a joyous multitude who strewed in the way these same graceful, softly-rustling fronds'.[40]

In the second half of the century, and especially after the opening of the coastal railway in 1872, many British, French and German visitors became seasonal or permanent residents on the Riviera. The climate was ideal for northern Europeans in search of a winter health resort and the new kingdom of Italy was relatively stable politically. Those attracted include the French architect Charles Garnier, who built his own large villa on the coast (1871–3) with a tower with views of the sea, the coast, the old town of Bordighera and extensive groves of palms. Many settlers were keen to establish lush gardens with tropical overtones. Perhaps the most important of these gardeners were the Quakers Thomas Hanbury (1832–1907) and his brother Daniel (1825–1875). Thomas Hanbury was a very wealthy tea, silk and cotton merchant at Shanghai, while his brother Daniel was a pharmacologist and botanist in London. Their first step was 'to increase the natural vegetation on the wild parts of the property, then almost denuded by the unchecked depredations of the neighbouring peasants, who had freely cut the trees and brushwood, and pastured their goats on the scanty herbage.'[41] They then started to establish a subtropical botanical garden covering 18 hectares. One of their gardeners was the German landscape architect and entrepreneur Ludwig Winter, who settled in Bordighera in the 1870s and established his own nurseries and started planting palms in Val di Sasso. Winter's gardens themselves became a tourist attraction and were advertised in the local international newspaper, the *Journal de Bordighera*, as 'Scheffel's Palms, near Madonna Della Ruota', where a 'beautiful group of palms stands in a fine garden on the promontory ... For cards of admission apply to Sig. L. Winter Bordighera.'[42]

The Scheffel palms were named after the popular German poet Joseph Victor von Scheffel (1826–1886), who had been fascinated by the palms on a visit in 1853 and described them in his influential poem 'Near Death' in 1856. This poem long remained popular: the German Nobel Prize-winning chemist Wilhelm Ostwald (1853–1932), following a nervous breakdown in 1895, 'travelled to the Mediterranean to get as much sun as possible. I stayed in Bordighera just because of the "seven palms by the seashore" mentioned in J. V. Scheffel's expressive poem.' He attempted to use painting as therapy and recounted that on his visit to Bordighera he kept to a strict routine 'of getting up early and going off on lonely walks with my paint box.'[43] The Val di Sasso near Bordighera became an iconic landscape and many artists depicted the palm groves and Ludwig Winter's garden. A large oil painting of the Val di Sasso (1888) by the German artist Hermann Nestel (1858–1905) was widely disseminated through prints; he also painted a fresco of the valley for the luxurious Hotel Angst in Bordighera.

The increasing popularity of palms and the rise of tourism caused local authorities to consider planting street trees and avenues so that visitors received a suitable impression of wealth and health. The popularity of Nervi, near Genoa, as a health resort increased once the Genoa to La Spezia railway opened in 1874. A new road was built to link the railway station to the town centre. The town council appointed a commission of three experts to design the new plantings and *stupendi palmizi* (beautiful palm trees) and orange trees were planted. The Viale delle Palme, which remains today, provided an impressive entrance to the resort from the railway station.[44] A few years later a similar avenue leading up from the railway station was established at Bordighera. The *Journal de Bordighera* of 3 November 1898 noted, 'We are glad to hear that there is some idea of planting it with palms on either side.' This would 'add materially to the picturesque effect of the road and would not fail to give visitors on their first arrival at the railway station, a favourable impression of the beauty of our town'.

Pierre-Auguste Renoir visited Bordighera briefly on his way to Venice in 1881, and in December 1883 he travelled with Claude Monet (1840–1926) by train along the coast to Genoa and back to Antibes. Monet stayed at Bordighera for three months in 1884 and many of the paintings he worked on there include palms set in villa gardens (illus. 45). He was particularly delighted to gain access to the private garden owned by Francesco Moreno, which according to some guide-books 'had more palms and other subtropical species' than any in Europe. He was enamoured of the 'lavish ochres of the palm fronds' and told friends in Paris that 'These palms are driving me crazy; the motifs are extremely difficult to seize, to put on canvas; it's so bushy everywhere, although delightful to the eye.'[45]

By the early twentieth century memories of traditional palm plantations and cultivation for ecclesiastical use had largely disappeared. The palm was firmly established as a symbol of luxurious gardening and wealthy health tourism. Palms are still planted and in the redeveloped harbourside at Genoa mature trees form a key element of the tourist experience. But many trees are now threatened by the red palm weevil, known in Italian as Punteruolo rosso (*Rhynchophorus ferrugineus*). This highly invasive pest is native to southern Asia but spread rapidly in the 1980s and from 2000 onwards reached many parts of the Mediterranean. Infestation often causes the palm tree to die and many mature palms in danger of collapse have had to be felled and removed. Palm trees that for many years were well-established features of railway stations, as at Rapallo, have been felled. More positively, some fragments of the old traditional palm plantations survive in places like the Val di Sasso, and as semi-wild populations in the river valleys at Taggia and Nervi. When I visited the Val di Sasso in 2018 with the geographer Pietro Piana we could see that, although traditional palm management had been abandoned, there were many surviving palm trees along the valley (illus. 23). Several terraces are now over-grown with vegetation such as broom (*Genista* spp.) and tree heather (*Erica arborea*). Semi-spontaneous plantations of the palm *Chamaerops humilis*, with scattered specimens of *Phoenix canariensis*, are found along the terraces. Palm trees are often surrounded by olive trees, iso-lated eucalyptuses and clumps of mimosa. The eucalyptus has been grown since the 1970s for decorative foliage.

The historical fame of the Val di Sasso survives in the memory of some local inhabitants: in an interview, the novelist Italo Calvino's gardener Guglielmi Libereso recalls the palms of the Val di Sasso, which reminded him of the landscape of an oasis in Africa, stressing the importance of the water canal for the cultivation of the trees.[46] Conservation schemes have been established to protect and publicize the history and conservation of palms and restore footpaths along the valley so celebrated by nineteenth-century authors. These restored paths provide a rare route for tourists to leave the beaches and visit and enjoy this once famous but now largely forgotten and gently decaying exotic landscape.

Street trees

The planting of trees for shade and beauty alongside streets in towns and cities has long been encouraged. The keen horticulturalist John Evelyn, writing in 1670, promoted the planting of lime trees in towns, wondering whether there was 'a more ravishing, or delightful object than to behold some intire *streets,* and whole *Towns* planted with these *Trees,* in even lines before their doors, so as they seem like

37 Peter Paul Rubens, *The Calydonian Boar Hunt*, c. 1611–12, oil on panel.

38 Samuel Howitt (1756–1822), *A Boar and a Ram*, n.d., watercolour.

The Well, Well Walk, Hampstead

39 Trees growing in Well Walk, Hampstead, *c.* 1918, postcard.

40 Galls of the oriental chestnut gall wasp
(*Dryocosmus kuriphilus*), Valletti, Liguria, 2014.

41 Joseph Keppler, *The Deadly Upas Tree of Wall Street*, 1882, chromolithograph.

42 Roger Fry, *E. M. Forster*, 1911, oil on canvas.

43 Ascan Lutteroth, *Scheffel's Palms*, 1883, oil on canvas: detail showing tied palms.

44 Henry Alford, *Behind Bordighera*, 1870, chromolithograph.

45 Claude Monet, *Garden at Bordighera, Morning*, 1884, oil on canvas.

46 Charles Mottram, after John Martin,
The Plains of Heaven, 1857, hand-coloured mezzotint.

Cities in a *Wood?*' He thought they were good for the health of residents and could have an 'admirable effect against the *Epilepsie,* for which the delicately scented *blossoms* are held prevalent'. Moreover, the trees helped to screen houses from '*Winds, Sun,* and *Dust;* than which there can be nothing more desirable where Streets are much frequented'.[47] But trees are often difficult to establish and frequently vandalized or die from lack of adequate management. An avenue of oak trees was planted by a Nottingham corporation in 1850 as part of a large town planning scheme which included the construction of an arboretum. A local councillor, William Parsons, noted in his diary on 14 March 1850 that he and his son had planted a tree in this avenue. But a few months later (21 November 1850) he met 'Mr H M Wood the Corporation Surveyor and his workmen in the Corporation Oaks Avenue and planted a new Oak Tree in the place where Fred and I had set one on the 11th Febry last but which with 13 others then planted had died'. He was pleased that the new tree was 'to bear the name of Parsons' Oak'.[48]

The planting and management of street trees has often caused controversy. Who owns the trees? Who is responsible for their protection? Who is liable for any damage they might cause? The historian Paul Elliott has shown how these debates played out in burgeoning British industrial cities of the late nineteenth and early twentieth centuries. In Glasgow in 1896 the parks committee of the new city corporation took control of the care of all street trees, planting new trees and removing 'old, decaying or dangerous' ones. The streets were the routes for increasing numbers of services, including water mains, surface water drains, telegraph and electricity supplies, sewers and gas pipes, which threatened tree roots. Trees were damaged by novel communication equipment: in 1901 trees along Pollockshaws Road were injured by 'telephone department workmen laying cables'. In Cardiff the corporation had looked after street trees planted by private developers since 1871 and started to plant its own trees from the 1880s. But there was concern that trees were planted to beautify streets in wealthy neighbourhoods and not poorer ones. It was argued in 1889 that 'it would be improper to fix on the poor people of the town the expense necessary to beautify and adorn the aristocratic quarters.' The enthusiasm for tree planting was such, however, that a proposal to stop public money being spent on 'the planting and guarding of trees' was defeated by 28 votes to 3.[49]

A particularly contentious furore over street trees took place in Sheffield between 2012 and 2018. The 2022 film *The Felling: An Epic Tale of People Power* is 'a first-hand account of an extraordinary and shocking tale, where Sheffield citizens from 18–90 years old rise up and become modern day heroes by putting their bodies on the line to stop the destruction of their city's healthy street trees and help save the Planet from climate Armageddon'.[50] There were violent disputes and arrests, which resulted in a great deal of adverse publicity for the city. Sheffield was a most unlikely setting for such an acrimonious debate. For decades Sheffield City Council had been well known for supporting environmental issues and local initiatives in tree and woodland management. The city was proud of its heritage of around eighty ancient woods and celebrated the street trees which were 'the key to forging, strengthening and creating "green corridors"'.[51] The background to the Sheffield dispute is a 25-year agreement between the City Council, the Department of Transport and Amey PLC to improve Sheffield's roads entitled 'Streets Ahead'. This agreement stated that the company 'shall replace the highways trees in accordance with the annual tree management programme at a rate of not less than 200 per year so that 17,500 highway trees are replaced by the end of the term'. A small number of the trees concerned were 'dangerous, dead, diseased or dying' but there were two further categories: 'damaging' and 'discriminatory'. These two categories included trees that were deemed to be damaging roads and pavements. Some trees were identified as discriminatory because 'they prevented people passing by in wheelchairs or when pushing prams.'[52]

Once the plans for tree felling became known various groups, together known as STAGS (Sheffield Tree Action Groups), were established to fight the proposals, the policy was challenged and alternatives were suggested: 'From 2015, until the tree-felling programme was "paused" in March 2018, people in Sheffield carried out "nonviolent direct action" to prevent the unnecessary felling of healthy street trees.' Simon Crump and Calvin Payne 'both received suspended prison sentences for breaching a High Court injunction' brought by Sheffield City Council (scc) 'to prevent peaceful direct action by protesters seeking to stop the felling of healthy street trees.'[53]

The testimonies of those involved in the dispute provide telling insights into people's attitudes to individual trees. Elizabeth Gash-Wales of Chippinghouse Road was very concerned when she received a letter in December 2016 'about the felling of Lonesome George', a large horse chestnut which she thought of as 'my tree ... How could they take such a beautiful, healthy tree?', which produced 'hundreds, possibly thousands, of conkers every year'. Her children had 'grown up with that tree. It's been a source of fascination and fun for years.' 'Lonesome George was one of the first things' her daughter saw when she was born. The tree was a 'source of many things for my children ... and we always ended up with bags full of lush, shiny conkers, which they'd take to school to share out'. Following the letter, 'I decided to use up two weeks of my annual leave to try and save him and other trees further up the road. I was out daily, like many others, standing in the cold.' However, her attempts, together with those of her friends and fellow-protesters, to protect Lonesome George were unsuccessful. When the tree fellers arrived 'we were called "stupid cows" by one of the Council workers.' The crowds were controlled by police: there were 'vans-full, on every corner and hiding on other streets'. The tree was felled. Elizabeth noted that the 'day we all stood, watching my tree being cut down so brutally (and with glee by certain arbs) remains with me today. I cried as people drifted away and sobbed later on my own.' For a while 'Lonesome George's trunk was left, standing like a beacon of all the spite and malice scc and Amey had been throwing at us.' But the 'glimmer of hope' that it would be left and allowed to grow on as a pollarded tree was soon dashed when it was seen 'being loaded onto the back of a truck'. She 'honestly felt bereft, almost like I'd lost a friend, and I suppose I had'.[54]

For some the tree campaign 'took over every day, it was on our doorsteps, every morning, every evening, every night. It was all-encompassing and exhausting.' These deep feelings for the protection of trees were not felt by all residents. Some were concerned about the damaging effect of the long dispute on the community: 'There are people I no longer speak to, people I've known for years. Neighbours that I used to smile at and say "Hello" to. I don't anymore, and I'm sure that's mutual.' On Lismore Road there were a few people 'on the street who were opposed, probably five of us, actively, *really* against it on this road, but most other people were sitting on the fence or actually in favour of it'. One neighbour 'had been asking the Council to take his tree down for fifteen years and he kept saying to us, "You're not going to stop me having my tree down, are you?", and I said, "Well we won't stop you having your tree down as long as you don't stop us having our tree up."' On Vainor Road one of the first trees scheduled to be felled was one that was categorized as dangerous and the closest resident 'was quite aggressive about it coming down'. Indeed 'the tree did have a big lean on it' and 'If you went to bed every night afraid that a tree was going to fall on your house, well, that's not great.' The campaigners did not contest the felling of that tree.[55]

In September 2017 anti-felling campaigners were receiving support from the *Daily Mail* and the Secretary of State for the Environment, Michael Gove, who told the *Yorkshire Post*: 'Sheffield is losing, we are losing, an amazingly valuable natural resource and the justification for it seems as flimsy as an autumn leaf.' He argued that the idea that pavements and kerbstones damaged by tree roots 'might theoretically pose a risk to someone's mobility and therefore justifies felling trees that have been here for generations is bonkers'.[56] Eventually, following 'mediated talks' between Sheffield City Council, STAG and Amey in 2018 and extensive public consultation, the *Sheffield Street Tree*

Partnership Strategy, designed to promote and enhance 'a network of street trees that Sheffield can be proud of', was published in May 2021.[57]

An example of the improved fate for Sheffield street trees is provided by 41 trees memorializing pupils of Western Road School. The trees were planted in 1919 and a nearby stone states 'the trees in Western Road and Gillott Street were planted in grateful appreciation of the part taken by former pupils of this school in the Great War 1914–1919.'[58] It is likely that the hard evidence of this finely inscribed monument helped to save the trees. Even so, their fate was in the balance for several years. In December 2017 Sheffield City Council decided that 'the option to retain the damaging trees would be a significant cost to the Council and provides only a partial and potentially short term solution.' The total estimated cost was around £500,000. They concluded that if the trees were not felled the damage they were likely to cause in the future 'will almost certainly give rise to claims against the Council or Amey and may make houses uninsurable'. By 2021, however, a compromise was reached which allowed the memorial trees to be preserved 'as well as upgrading both the road and pavement surface'. This solution included the use of 'flexi-pave', made partly from recycled rubber tyres, which 'enables tree roots to grow and move, without damaging them or the outer surface'. It was planned to use this technical solution in other parts of the city with 'similar design challenges'.[59]

A Labour councillor interviewed in 2017 at the height of the dispute was strongly in favour of the tree felling and replacement programme. He argued that the protesters were a few people who 'are causing a lot of upset and are wasting a lot of money. Most people in the city, the people I represent, are just trying to get on with their lives.' He also emphasized the importance of class divisions within the city. The people complaining were from 'the leafy suburbs, the west of the city trying to dictate what should be done'. They were wealthy and can 'attract attention but they're not elected'. He saw the dispute as 'a bloody class war! This is a divided city. Always has been but it's getting worse.' The people to the west 'loving their trees and all that sort of thing while the rest of the city have to get on with real life'. He stressed the impact of 'massive, I mean bloody massive, cutbacks' and that 'We're in a time of tough choices about social care, housing, asylum seekers, vulnerable people ... people on the streets and desperate ... and then we have the trees.'[60] These divisions were of long standing and older than the threatened trees. John Betjeman in his poem 'An Edwardian Sunday, Broomhill, Sheffield' celebrates the extensive gardens of Sheffield's western suburbs: 'A sylvan expansion/ So varied and jolly/ Where laurel and holly/ Commingle their greens.' Here 'in our arboreta' the industrialists and merchants could look down from the heights of Broomhill, 'On back street and alley/ And chemical valley/ Laid out in the light' where wealth was produced.

7

SACRED TREES

William Wordsworth's sublime, horrific poem 'The Thorn' (1798) opens: 'There is a Thorn – it looks so old,/ In truth, you'd find it hard to say/ How it could ever have been young,/ It looks so old and grey'; it is overgrown with lichens. Yet the hawthorn is also strongly associated with youth, springtime and the first day of May. It is celebrated for its huge quantities of spring blossom and, indeed, named May after its flowering month. Marcel Proust famously remembers childhood walks with his grandfather along the path he names *Swann's Way* and recollects the scent and extraordinary beauty of the creamy hawthorn blossoms, frequently tinged with pink. The association between hawthorn and late spring is so strong that Simone de Beauvoir used the simile 'as startling and as poetical as finding a hawthorn in flower in the middle of winter' in her *Memoirs of a Dutiful Daughter*.

One tree, the Glastonbury thorn (*Crataegus monogyna* 'Biflora'), is famous for overturning and upsetting the close relationship between the May flowering hawthorn and the calendar. Its legendary association with Joseph of Arimathea and the propagation of Christianity in Britain has flourished for hundreds of years.[1] The common hawthorn, of which this is a form, is a deciduous, spiny shrub or small tree, whose natural distribution is in Europe, North Africa and western Asia. Its spines make it an ideal species for hedges, for which it has been used for centuries; it has been introduced to North America, where it can be invasive. There are many different horticultural varieties with differing flower colour and leaf form. The unusual form known as the Glastonbury thorn has a peculiarly anachronistic flowering habit, often flowering twice in the same year, and sometimes holds its leaves through the winter. These characteristics have encouraged it to become desired, appropriated and celebrated over many centuries by those with diverse religious, spiritual and historical interests and concerns (illus. 25).

The Glastonbury thorn can be understood in a broader context of meaning and place identification. The historical geographer Della Hooke and others working on Anglo-Saxon trees in England have shown that thorns are the most frequently mentioned tree species used for boundary markers, followed closely by oak and ash, in Anglo-Saxon charters and early English place names. Vaughan Cornish in his *Historic Thorn Trees in the British Isles*, published in 1941, investigated their distribution using a variety of historical sources, maps and fieldwork. He was drawn to his study by inheriting the Salcombe Regis thorn, near Sidmouth in Devon, 'which had been maintained from time immemorial in its original site by replanting'. He notes that as the hawthorn does not always live long, 'both my grandfather and my elder brother had to plant a Thorn in the time of their ownership.' The tree was blown down in 1928 and the 'District Council desired to take the opportunity of rounding off the sharp corner of the road where it had stood', so the 'present Thorn Tree is a few yards south-west of the original position'. When the thorn came into his ownership in 1938 he 'put up a monumental stone to tell the passer-by the history of the tree' and help preserve the memory of its history and legend. The plan worked, and the tree remains celebrated as one of the 'Great Trees of East Devon'.[2]

The hawthorn has many horticultural varieties selected and propagated over centuries. Loudon reported that there were many varieties of the common hawthorn and 'some of them very distinct'. He thought this was because the plant had been 'very extensively raised from seed, for making hedges; and curious nurserymen, when they have observed any plants indicating a striking peculiarity of foliage, or mode of growth, in their seed beds, have marked them, kept them apart, and propagated them by budding or grafting.' He noted over thirty varieties available from the principal nurseries, one of which was the Glastonbury thorn, which 'comes into leaf in January or February, and sometimes even in autumn; so that occasionally, in mild seasons, it may be in flower on Christmas-day.' Loudon recounted that 'according to Romish legend' there was still a thorn 'said to be a descendant of the tree' which had once 'formed the staff of Joseph of Arimathea' to be found 'within the precincts of the ancient abbey of Glastonbury'. This tree 'maintains the habitat of flowering in the winter, which the legend attributes to its supposed parent'. Loudon also reports on other examples sent to him, including one from Oxford Botanic Garden: the curator Mr Baxter sent a specimen 'gathered in that garden on Christmas-day, 1834, with fully expanded flowers and ripe fruit on the same branch'. But Loudon scoffs at the legend: 'there is nothing miraculous in the circumstances of a staff, supposing it to be of hawthorn, having, when stuck in the ground, taken root, and become a tree; as it is well known that the hawthorn grows from stakes and truncheons.'[3]

Such practical horticulture allows for continuity and reinforces origin myths associated with the first Glastonbury thorn growing from St Joseph of Arimathea's staff. Over the years different Christian denominations have captured, memorialized, labelled and fixed the position of trees through marker stones, maps and written texts and descriptions. One of the best-known Glastonbury thorns, grafted from a tree in St John's Churchyard, was planted on Wirral Hill by the Mayor of Glastonbury in 1951. This tree was attacked and had its branches cut off in 2010, but it remained 'maimed and limbless but much venerated from 2010 until 2019 when the trunk itself was removed'. The individual ephemerality of trees is combatted by taking cuttings and replanting.[4]

Cedars in Lebanon

The increasing numbers of tourists to the Holy Land in the late nineteenth century elicited fascination with the fate of the cedars of Lebanon. With the increasing popularity of landscape photography, images of the cedars growing in Lebanon began to be used as illustrations for books, including the Bible. One of the most important landscape photographers was Francis Frith (1822–1898), who turned his early amateur enthusiasm in photography into a profitable business. He made his reputation with photographs taken on journeys to the Near East including Palestine and Syria in the 1850s, 'where he took pioneering photographs of the landscape and of monuments, often under dangerous and difficult conditions.'[5] Photographs sold by the company included general views of the cedar groves and photographs of individual trees (illus. 47).

Visitors were concerned that heavy grazing by sheep and goats meant that there was hardly any natural regeneration of the cedars (*Cedrus libani*). A 'detailed survey of the basin where the cedars grow' at the head of the Kedisha valley at 1,800 metres had been made by two Royal Navy surveyors with Sir Joseph Hooker in 1860. Hooker 'believed that the wood used by Solomon and by Nebuchadnezzar in buildings was the Lebanon cedar'. The age of the cedars was estimated by making ring counts from a branch of an old tree. The youngest trees were about 100 years old and the oldest 2,500 years old. The 'most remarkable and significant fact connected with their size, and consequently with the age of the grove, is that there is no tree of less than 18 inches girth, and that no young trees, seedlings, or even bushes of a second year's growth were found.'[6] In other words, there was no successful natural regeneration, and the future of the cedar groves was in peril.

It was reported in the *Gardeners' Chronicle* (1879) that 'for want of proper protection against the goats and thoughtless tourists, the present grove is dwindling away, and another generation will exclaim against our supineness in thus allowing a relic of the past to die out prematurely.' A visitor in 1903 reported that a grove he had visited 'suffers much from being cut' and he noticed that 'Local people obtain from it roof-beams and wood for fuel . . . I have failed to find a single large tree . . . which has not been cut off, with the result that several branches have taken the place of the principal stem.' There is a remarkable photograph taken by Cornelius Van Alen Van Dyck (1818–1895), an American doctor, Protestant missionary and translator, showing a splendidly isolated cedar sheltering a sheep and goats, in a scene which temptingly takes us back to Old Testament times.[7] Many cedars in Lebanon were felled during the First World War as fuel for railways. Cedars were often coppiced 'at the sapling stage, which produces multiple-stemmed trees', especially where trees were growing close to villages such as Hadeth and Barouk. A study of remnant cedar trees at Barouk found that the older trees 'often consisted of two or three boles branching near the ground', and that when dated with an increment borer 'five large trees at Barouk gave ages from 375 to 440 years.'[8]

47 Francis Frith, *Cedars of Lebanon*, c. 1857, photograph.

While the numbers of cedars growing in Lebanon itself have decreased over the last five hundred years, the species has become one of the most popular trees to be planted in parks and gardens around the world. A lot of the interest in the tree derived from its biblical associations. The cedar of Lebanon is mentioned in some of the most vivid stories in the Bible. At the end of the sixteenth century the herbalist John Gerard described the '*Cedrus Libani . . .* the great Cedar tree of Libanus' as 'huge and mightie' and as having branches 'so orderly placed by degrees, as that a man may climb up by them to the top as by a ladder'. He notes that the 'Cedar trees grow upon the snowie mountains, as in Syria on mount Libanus, on which there remaine some even to this day . . . planted as it is thought by Salomon himself.'[9] Early French and British travellers to Syria and Lebanon visited Mount Lebanon and were fascinated by the idea that the individual trees they saw were the same as those described by biblical prophets. The French naturalist Pierre Belon (1517–1564), who published an account of his travels through the Near East in 1553, was like most travellers guided to the cedars on Mount Lebanon by the Maronite Eastern Catholic monks from the monastery of the Virgin Mary.

Loudon noted that at this period 'paying a visit to the cedars of Mount Lebanon seems to have been considered a type of pilgrimage'.[10] But the pilgrims were accused of damaging the trees as they 'took away some of the wood of the trees, to make crosses and tabernacles'. The damage was so great that 'the patriarch of the Maronites, fearing that the trees would be destroyed, threatened excommunication to all those that would injure the cedars' and 'exhorted all Christians to preserve trees so celebrated' in the Bible. It was reported that 'the Maronites were only allowed to cut even the branches of these trees once a year: and that was on the eve of the Transfiguration of our Saviour,' which is in August and 'a suitable period for visiting the mountain'. At this festival the 'Maronites and pilgrims' climbed the mountain and, 'passing the night in the wood', drank 'wine made from grapes grown on the mountain, and lighted their fires with branches cut from the cedars. They passed the night in dancing a kind of Pyrrhic dance, and in singing and regaling,' and on the next day 'the patriarch celebrated high mass on an altar built under one of the largest and oldest cedars.'[11]

One of the earliest surviving first-hand English descriptions of the cedars was provided by Henry Maundrell, chaplain to the Levant Company's factory at Aleppo. He travelled to the cedars on 9 May 1697 and found 'noble trees' growing 'amongst the snow, near the highest part of Lebanon'. He thought the trees 'remarkable as well for their own age and largeness, as for the frequent allusions made to them in the Word of God'. There were only sixteen 'very old trees, of a prodigious bulk', but there were 'numerous' younger trees. He measured 'one of the largest, and found it 12 yards 6 in. in girt, and yet sound; and 37 yards in the spread of its boughs'.[12] About forty years later Richard Pococke, later Bishop of Ossory, found the 'famous cedars of Lebanon' formed 'a grove about a mile in circumference, which consists of some large cedars that are near to one another, a great number of small cedars, and some young pines'. He also reported that 'Christians of several denominations near this place come here to celebrate the festival of the Transfiguration, and have built altars against several of the large trees, where they administer the sacrament.'[13]

John Evelyn in the 1670 edition of his *Sylva* uses the reported small number of surviving cedars on Mount Lebanon to argue for improved tree management. He notes that the 'Cedar in Judea was first planted there by Solomon, who doubtless try'd many rare Experiments of this nature; and none more Kingly than that of Planting to Posterity'. The clear implication here is that his readers should follow Solomon's example and plant trees for the benefit of future generations. He then refers to travellers' descriptions which indicate that only a few of these biblical trees survive and argues that this is 'a pregnant Example of what Time, and Neglect will bring to ruine, if due, and continual care be not taken to propagate Timber'.[14] This clear link between cedars and Lebanon becomes blurred, however, when he considers the quality of cedar wood. He notes that the cedar 'grows in all extreams: In the moist Barbados, the hot Bermudas, the cold New-England; even where the Snow lyes as (I am assur'd) almost half the year'.[15] And although he hopes fancifully that the use of timber by London merchants in their shops might 'preserve the whole City as if it stood amongst . . . the prospects of Mount Libanus', he is using the term 'cedar' for many different timbers and indeed considers some juniper wood as 'sweet as Cedar whereof it is accounted a spurious kind'.[16]

But in the third edition of his book (1679) he mentions that he has obtained seeds from Mount Lebanon and he is therefore able to describe at first hand what the cones looked like. He had 'received Cones and Seeds of those few remaining Trees' from 'on the Mountains of Libanus' and, as they grew there, he questions, 'Why then should they not thrive in Old England?' and argues, 'I know not, save for want of Industry and Trial'.[17] His enthusiasm for the tree is clear in his precise description of the cones, which he handled himself: 'These Cones have the Bases rounder, shorter, or rather thicker, and with blunter Points, the whole circum-zoned, as it were, with pretty broad thick Scales, which adhere together in exact Series to the very Top and Summit, where they are somewhat smaller.' But he also reports continued uncertainty about the status of the tree: 'Botanists are not fully agreed to what Species many noble and stately Trees, passing under the Names of Cedar, are to be reckoned.'[18]

The cedar of Lebanon today has a limited natural distribution in the Eastern Mediterranean. The area of cedar woodland has been greatly reduced over the last 5,000 years through tree felling and grazing by goats.[19] There is great concern about the potential impact of climate change, warfare and the depredations of the cedar web-spinning sawfly, or *Cephalcia tannourinensis*.[20] Although there are a few remaining cedar outliers in Lebanon itself, the largest area of surviving woodland is in the Taurus mountains of southern Turkey.[21] Here there is optimism that with specialized forestry techniques, including the use of natural regeneration and prescribed burning, vigorous new generations of cedars can be established in existing woodland. There has, moreover, been considerable success in the plantations of cedar of Lebanon on bare karst landscapes or former oak woodland.[22]

Sacred cedars in England

No one knows who first grew cedars of Lebanon successfully in England.[23] It is possible that it was John Evelyn and Loudon argued that

> It is extremely improbable that a man so fond of trees as Evelyn, and so anxious to introduce new and valuable sorts into his native country, should have suffered 'cones and seeds' of such a tree in his possession, without trying to raise young plants from them; particularly as he was a man of leisure, residing in the country, and fond of trying experiments.[24]

Another strong contender is the oriental and biblical scholar Edward Pococke (1604–1691), who held the first chair in Arabic at Oxford in 1636 and was also Regius Professor of Hebrew. He was chaplain to the Levant Company at Aleppo from 1630 until 1636, where he was able to study oriental languages and collect manuscripts. He was a tree enthusiast and several 'noble trees planted from seeds that Pococke brought back from the east commemorated him long after his death'. He was rector of Childrey in Hampshire and, 'according to unbroken tradition', a large cedar tree he planted in the rectory garden survives to this day.[25] Pococke pointed out that there were significant difficulties for the biblical scholar in the identification of plants and animals. He noted that the 'green fir tree' mentioned at Hosea 14:8 could in different Arabic and Hebrew texts be translated as 'a leafy juniper tree', 'a fertile juniper tree' or 'a kind of cedar'. He argued 'That there should be a difference and ambiguity in rendring the names of plants or animals among such who lived not in the place where those things were, and saw and heard what was so or so called, is no wonder.'[26]

By the mid-eighteenth century the fashion for planting cedars of Lebanon was well established among the British aristocracy. One of the greatest enthusiasts for introducing new, exotic trees was the 3rd Duke of Argyll, who was known as the 'tree-monger of Whitton'; his estate on Hounslow Heath had a vast collection of introduced trees and a nursery. The duke started his collection in the early 1720s and trialled many introduced trees to see if they could withstand the English climate. There was 'a very large number of Cedars of Lebanon.'[27] Cedars were planted with enthusiasm by both Lancelot 'Capability' Brown, as at his first major commission at Croome Park, Worcester-shire, in the 1750s for the 6th Earl of Coventry, and by his Picturesque critic Uvedale Price at Foxley, Herefordshire, later in the century.[28] At Croome, the Rotunda, a circular garden temple designed by Capability Brown and Sanderson Miller, was built about 1754. Cedars of Lebanon circle the temple, encompassing the landscape views that can be seen from its six windows. Estate records show that an expen-sive 8-foot-high cedar of Lebanon had been purchased by the Earl of Coventry for two guineas in 1748; twelve more were bought in 1749, with others in following years. The trees surrounding the Rotunda were planted in sympathy with the building and cedars are now iconic features of the Croome landscape (illus. 24).[29]

The Picturesque enthusiast William Gilpin, vicar of Boldre in the New Forest, argued in 1791 that the cedar of Lebanon was pre-eminent among all evergreens not only because of its own 'dignity', but 'on account of the respectable mention, which is everywhere made of it in scripture'. He emphasizes that 'Solomon spake of trees from the cedar of Lebanon, to the hyssop that springeth out of the wall: that is from the greatest to the least.' Moreover, the 'strength of the cedar is used as an emblem to express the power even of Jehovah.' He thought that Ezekiel identified 'with great beauty, and aptness' the 'shadowing

48 Henry Warren, after, James Fuge, *North View of the Cedar Trees in the Garden of the Apothecary's Company, Chelsea, c.* 1850, lithograph.

shroud', as 'no tree in the forest is more remarkable than the cedar' with its 'close-woven, leafy canopy'. It was this 'mantling foliage', arising from 'the horizontal growth' of branches forming 'a kind of sweeping, irregular penthouse', that gave the cedar its 'greatest beauty'.[30]

Gilpin was not convinced that the British climate suited the establishment of large numbers of cedars as old cedars frequently become 'shrivelled, deformed, and stunted'. There were exceptions, however, including a well-known tree at Hillingdon, near Uxbridge, which when he saw it in 1776 'was about one hundred and eighteen years of age: and being then completely clump-headed, it was a very noble and picturesque tree.' This cedar tree at Hillingdon had become one of the most famous in England, clearly visible from the main road. It had been planted by Samuel Reynardson, who lived there from 1678 to 1721. Gilpin thought it the 'best specimen of this tree I ever saw in England' and measured it: the 'perpendicular height of it was 53 ft., its horizontal expanse 96 ft., and its girt 15 ft. 6 in. . . . this noble cedar was blown down' in 'the high winds about the beginning of the year 1790'; 'its stem, when cut, was 5 ft. in diameter.' Sir Joseph Banks had a table made from its timber.[31]

Other famous old cedars were drawn and described by Jacob George Strutt (1784–1867), who thought that 'The frequent and solemn allusions to the Cedar of Holy Writ, seem to give it something of a sacred character,' which was reinforced by 'knowledge of the esteem in which it was held by the ancients'. Some individual trees, such as the 'great cedar at Hammersmith', when 'in the full prime of its summer foliage, waving its rich green arms to the gentle breezes, and hiding the small birds innumerable in its boughs', afforded 'a fine exemplification of the sublime description of the Prophet Ezekiel, in his comparison of the glory of Assyria, in her "most high and palmy state".' Strutt's etching shows women and children taking advantage of the shade cast by the tree. Others were growing less well, including those in Chelsea Physic Garden, apparently planted in 1683, whose branches have 'of late years altogether drooped and languished' owing to the 'pestiferous vapour of the numerous gas-works by which it is surrounded' (illus. 48).[32]

Concern about the impact of air pollution on cedars was reiterated throughout the early years of the twentieth century. W. J. Bean, curator at the Royal Botanic Gardens at Kew, had no doubt that

> Irrespective of its sacred and historical associations, no
> tree ever introduced to our islands has added more for
> the charm of gardens than the cedar of Lebanon. Its
> thick, stately trunk and noble crown of wide-spreading,
> horizontal branches give to it an air of distinction no
> other tree at present can rival.

But he was disturbed by the impact of industrial and domestic pollution and the intensity and frequency of smog, which meant that cedars growing in London's suburbs were 'becoming fewer and less vigorous'.

He thought that 'until there is a revolution in the methods of consuming coal in the metropolis, the gaps will never be filled.'[33]

It was recognized that timber produced by the cedar of Lebanon in England was not of a particularly high quality. Loudon minimized its commercial value: 'The wood of the cedar is of a reddish white, light and spongy, easily worked, but very apt to shrink and warp, and by no means durable.' Authors found it difficult to equate the quality of the wood grown in France and England with the qualities described by classical authors and in the Bible. The table Sir Joseph Banks had made from the Hillingdon cedar 'was soft, without scent (except that of common deal), and possessed little variety of veining'. Moreover, cedar wood 'burns quickly, throwing out many sparks, though but little heat in comparison with that of the oak or the beech; though the flame of the cedar wood is more lively and brilliant, on account of the resin it contains.'[34] Earlier, Walter Nicol had noted that 'This celebrated tree is found in the highest perfection in the bleakest and most mountainous sites of the East; but whether it shall be found so on the mountains of Britain, remains to be known.'[35] Henry Elwes, writing in 1908, confirmed the low value of cedar wood in Britain: 'Its value in commerce is . . . low, because neither the supply nor the demand is regular; and the cost of removing and sawing up large cedar trees is so great, that I was offered a tree containing 300 cubic feet for nothing if I could get it away.'[36] The gardens of East End House, Hammersmith, once the residence of William IV's mistress Mrs Fitzherbert, 'contained a magnificent cedar-tree, which had to be blown up by dynamite when the house was pulled down' in 1885.[37]

By the Victorian period cedars of Lebanon had become a key emblem of country house gardens and parks. Loudon enthused that 'As an ornamental object, the cedar is one of the most magnificent of trees; uniting the grand with the picturesque, in a manner not equalled by any tree in Britain, either indigenous or introduced.'[38] He thought that if large numbers were to be planted and 'a distance of 50ft. or 60ft. allowed between each tree', then 'nothing in the way of sylvan majesty can be more sublime than such a forest of living pyramids.' William Ablett, in 1880, noted that the 'elegant grandeur of the cedar was often used as a type and illustration by the Hebrew prophets, to express the beautiful steadfastness and comely aspects of the spiritual condition of the righteous, which earthly storms, though they might shake, could not remove', while the laconic John Blenkarn noted that the 'beauty of this tree is so well known and acknowledged, that any comment here is unnecessary'.[39] Some owners were so keen to have a mature cedar that they could not wait for young trees to attain the desired age and form. At the 4th Earl of Harrington's Elvaston Castle near Derby, William Barron devised a tree transplanting machine and in November 1831 successfully moved a cedar of Lebanon 43 feet tall into the gardens at Elvaston that had grown by the 1870s from having a trunk of 2 feet in circumference to 10 feet.[40]

In addition to the cedar's association with the gardens and parks of large houses, they became strongly linked with sacred spaces such as churchyards and cemeteries. Loudon wrote his influential treatise *On the Laying Out, Planting, and Managing of Cemeteries; and on the Improvement of Churchyards* in 1843. He recommended cedars for new urban cemeteries and established country churchyards, but recognized that, as with many other conifers, its habit of surface rooting could cause problems. Indeed, he thought that if it were not for this 'the cedar of Lebanon would be one of the most fitting of all trees for a churchyard, from the sombre hue of its foliage, and its grand and yet picturesque form; from the horizontal lines of its spreading branches contrasting strongly with the perpendicular lines of a Gothic church' and its biblical associations. To get over this problem of surface rooting, he recommended that in all new churchyards 'two or three spots (each about 30 ft. in diameter) were set apart, not to be broken up by interments, and each planted with a cedar of Lebanon'. In addition, he felt that

in many old churchyards in the country, a spot sufficiently large for at least one cedar might easily be spared; and the clergyman or the churchwardens who should plant a cedar on such a spot, and fence it sufficiently while young, would confer a grand and appropriate ornament on the church, and would deserve the gratitude of the parishioners.[41]

One of the most famous examples of a new cemetery was at Abney Park, Stoke Newington in north London. This was laid out in 1840 with the help of the leading nurseryman George Loddiges and an immense collection of trees and shrubs was planted.[42] George Collison, one of its promoters, published a guidebook which extolled the beauty of the existing well-established trees including old cedars of Lebanon.[43] The most famous of these trees were those surrounding the former residence of the Congregationalist hymnologist Isaac Watts (1674–1748). About 1712 Sir Thomas Abney invited Watts to stay with him 'for what was initially a week's stay', but 'Watts was to spend the rest of his life in considerable comfort with the Abneys' at their various homes and from 1733 he stayed 'with Lady Abney and her two surviving daughters', who 'represented for him the ideal dissenting household' at Abney Park.[44] By the nineteenth century it was claimed that some of the older trees at Abney had been planted by Watts and his friends, especially 'the magnificent cedar of Lebanon'. As Isaac Watts had spent much of his life enjoying the grounds, park and trees at Abney House and had written some of his famous works there, the old trees were strongly associated, especially in the minds of dissenters, with his works. For promoters like Collison, these specific religious associations were merely part of the general appeal of the contemplation of trees at Abney:

> No true lover of nature needs to be reminded of the
> pleasure which the mind receives in the contemplation
> of trees, remarkable either for size or longevity – those
> silent, yet eloquent historians of the passing generations
> of the human race . . . all ages have stood indebted for their
> earliest and best-loved sanctuaries.[45]

As well as being planted to make particular places more sacred, representations of the trees were used by artists to characterize historical religious paintings and illustrations in sacred books. The painter and printmaker John Martin used the characteristic form of the cedar in several of his works, including his mezzotint *The Fall of Man* (1831), in which Adam and Eve are depicted having eaten of the tree of knowledge. Line-engraved copies of this print were very popular in the Victorian period and frequently used as illustrations to sacred works such as *The Home Preacher; or, Church in the House* (1868–9) and as the frontispiece to *The Imperial Family Bible* (1844).[46] A huge cedar, with cedars intermixed with other trees disappearing into the distance, dominates the right side of his extraordinarily sublime oil painting *The Plains of Heaven*, which formed the left panel of his *Last Judgement* triptych (*c.* 1845–53). This work 'effectively tapped into vernacular Christian culture' and was widely circulated as engravings and mezzotints (illus. 46).[47]

While the cedar of Lebanon remained threatened in its homeland, in England and many other countries it became a fashionable ornamental tree and a common specimen tree in large gardens and parks. But more than this, there was an explicit and overt attempt to establish the tree as sacred through planting in churchyards and cemeteries. The strength of the biblical connections was reinforced by the popularity of photographs of cedars of Lebanon, such as those by Francis Frith, in the later nineteenth century. Here photographs of ancient cedars, which were interpreted as direct, living connections with Old Testament characters, were used both to ground and place the Bible in the landscape.

Memorial pines

William Wordsworth was drawn to the idea of the fraternity of trees, referring in various poems to 'the brotherhood of venerable trees' and the 'brotherhood of lofty elms'. He once saw two trees on top of

Oker Hill near Matlock, Derbyshire, silhouetted against the sky, that were thought to have been planted by brothers; the trees 'now entwine their arms', but the brothers went separate ways and never met again.[48] William's brother John Wordsworth (1772–1805) was captain of the East India Company's *Earl of Abergavenny*, which foundered on rocks off Weymouth on 5 February 1805 with the loss of at least 250 lives.[49] The following spring William saw in London a painting by his friend Sir George Beaumont (1753–1827), *Peele Castle in a Storm*. He personally found this depiction of a wrecked ship on rocks below the castle to be 'a very moving one' and it inspired him to write his *Elegiac Stanzas Suggested by a Picture of Peele Castle in a Storm Painted by Sir George Beaumont* in 1806. Writing to Sir George about the poem Wordsworth said, 'It is a melancholy satisfaction to me to connect my dead Brother with any body whom I love so much; and I knew the verses would give you pleasure as proof of my affection for you.'[50] Even before his brother's death one of Wordsworth's favourite places was on the fell above Dove Cottage, where his walking back and forth created a 'hoary pathway' in a group of pine trees, which became known as 'John's Grove'. In 1802 Dorothy noted that she and William went 'to John's Grove, where we sate a little while looking at the fading landscape ... There was a sweet sea-like sound in the trees above our heads, we walked backwards and forwards some time for dear John's sake.' After John's death the pine grove became 'an almost sacred space.'[51]

Wordsworth had contrasting views of the Scots pine. On his visit to Scotland in 1803 he admired the trees and the 'gloom, and even grandeur' of a hillside near Glencoe 'thinly sprinkled' with them, which 'appeared to be the survivors of a large forest'. But in his *Guide to the Lakes* he disliked new plantations of dense, spindly 'Scotch Firs', which were 'less attractive' when young 'than any other plant'. He recognized that they could grow into 'noble' trees if given room to spread out their branches and was keen that 'aboriginal species' of Scots pine, 'which can only be procured from the Scotch nurseries', were to be planted rather than introduced American conifers.[52] But specific association with family and friends trumped aesthetics. This is demonstrated by his attitude to a particular stone pine (*Pinus pinea*) he visited on Monte Mario in Rome in 1837.

With Napoleon's defeat at Waterloo in June 1815, Rome once again became a favourite destination for British cultural tourists.[53] The poet Catherine Maria Fanshawe, writing on 11 January 1820 to her friend Lady Bury in London, emphasized her delight at being able to experience ancient and modern Rome:

Not all the preparation of traversing Sea and Land and
Mountains and Vallies quite accounts for the fact of
being actually at Rome and though I know it when
I lie down at night I am hardly less amazed when
I rewake to see the Pillar of Trajan before my window.

She was disappointed to report that she had missed her friend Sir Thomas Lawrence, president of the Royal Academy, by only four days, and was concerned that 'there is still here a swarm of English: not a very interesting set in general,' However, on 'the very night of my arrival I visited the Coliseum by moonlight and the next was shown the finest statues in the Vatican by torchlight, a treat indeed.' St Peter's 'afforded me the noblest pleasure': she felt she 'were beholding in Heaven the work not of angels but of beatified spirits decked in all the glory and magnificence of a celestial church'.[54]

The friends she met in Rome included Sir George and Lady Beaumont, who lived there from 1819 to 1822. The Beaumonts had a wide circle of literary and artistic friends including Wordsworth, Coleridge and Uvedale Price. Sir George was one of the leading connoisseurs, amateur artists and art collectors of the late eighteenth and early nineteenth centuries. His wealth, derived largely from his estates at Coleorton (Leicestershire) and Dunmow (Essex), allowed him to collect works by Rubens, Rembrandt, Claude and Poussin. Some of these he had purchased on his first visit to Rome in 1782–3. He was also a leading patron of John Constable. While in Rome, Catherine Fanshawe introduced the Beaumonts to Charles Eastlake, later director of the National Gallery to which Sir George donated his collection during his lifetime to help establish this new gallery. He also regularly met the sculptor Antonio Canova, who helped him to purchase Michelangelo's unfinished sculpture *Virgin and Child with the Infant St John*, also known as the Taddei Tondo, which is now at the Royal Academy.[55]

Sir George made excursions into the countryside with the poet Samuel Rogers, including one to Frascati 'through galleries or avenues of ilex and cypress', taking in views of Rome and the Campagna.[56] He particularly admired the stone pine trees of Rome and later told his friend Wordsworth that on his first visit in the 1780s 'pine-trees of this species abounded, but that on his return thither, which was more than thirty years after, they had disappeared from many places where he had been accustomed to admire them, and had become rare all over the country, especially in and about Rome'.[57]

Wordsworth visited Rome in 1837 with Henry Crabb Robinson.[58] Writing to his sister Dorothy in April 1837 he noted that 'we had scarcely been two hours in Rome when we walked up to the Pincian hill, near our hotel. The sun was just set, but the western sky glowed beautifully.' The 'modern' part of Rome was below them 'and St. Peter's rose on the opposite side'. He noted 'for dear Sir George Beaumont's sake' that 'at no great distance from the dome of the church on the line of the glowing horizon was seen one of those broadtopped pines, looking like a little cloud in the sky, with a slender stalk to connect it with its native earth.' The English sculptor William Theed, who worked in Rome, 'accosted' Robinson by name and told them that the tree which Wordsworth 'admired so much had been paid for by our dear friend' upon the condition 'that the proprietor should not act upon his known intention of cutting it down' so that 'it might stand as long as nature might allow'.[59]

A week later Wordsworth told his wife and sister that 'The Monte Mario commands the most magnificent view of modern Rome, the Tiber, and the surrounding country.' He climbed the hill and 'stood under the pine, redeemed by Sir G. Beaumont, of which I spoke in my former letter. I touched the bark of the magnificent tree, and I could almost have kissed it out of love for his memory.'[60] He later embellished the story, stating that 'having ascended the Monte Mario, I could not resist embracing the trunk of this interesting monument of my departed friend's feelings for the beauties of nature, and the power of that art which he loved so much, and in the practice of which he was so distinguished.'[61] The chance discovery of Sir George's pine tree was the stimulus for his sonnet 'The Pine of Monte Mario at Rome':

I saw far off the dark top of a Pine
Look like a cloud – a slender stem the tie
That bound it to its native earth – poised high
'Mid evening hues, along the horizon line,
Striving in peace each other to outshine.
But when I learned the Tree was living there,
Saved from the sordid axe by Beaumont's care,
Oh, what a gush of tenderness was mine!
The rescued Pine-tree, with its sky so bright
And cloud-like beauty, rich in thoughts of home,
Death-parted friends, and days too swift in flight,
Supplanted the whole majesty of Rome
(Then first apparent from the Pincian Height)
Crowned with St. Peter's everlasting Dome.[62]

The sonnet reinforced the importance of Roman pines for literary British visitors. Wordsworth reported that several Roman villas had in recent years 'passed into the hands of foreigners' whom he noticed had 'taken care to plant this tree, which in course of years will become a great ornament to the city and to the general landscape'.[63] Sir George Beaumont's action in saving a single, prominent pine which was a crucial part of a well-known landscape view is an early example of practical landscape preservation (illus. 26).

CONCLUSION

This book has explored the diverse interactions between people and trees over time. In broad terms there is agreement that the area of woody landscapes should be increased. This is because of deep public enjoyment of and interest in trees, woods and forests and the wide range of benefits provided by woodland. These include timber production, firewood, carbon sequestration, landscape and culture, wildlife and game conservation, and the increasing realization that public access and engagement with woodlands brings benefits for mental and physical health. These benefits make woodland a complex land use to understand and manage.

While there is general agreement on the values trees provide, the need to establish much more woodland and the best ways of managing it remain subjects of much debate, discussion and dispute. Those researching trees, woods and forests often come from cultural and scientific traditions that reinforce the positive attributes of trees. But people also have negative feelings about specific woody landscapes. Some woods can be associated with illegal, dangerous and anti-social behaviour. But these may be the same woods within urban, suburban and edgeland landscapes that are of enormous importance for people wishing to walk their dogs, or go jogging or mountain biking. They are places where children love to play, making dens, lighting fires, carving names on trees and scattering debris and litter: the ambiguously threatening type of landscapes which the artist George Shaw depicts in his acrylic paintings.[1] Trees can be dangerous, blocking site lines at road junctions, damaging houses and pavements, causing injury by falling branches and trees. The legal implications of occupiers' liability for dangerous trees are often a reason for felling trees and can discourage people establishing new ones.

The history of forestry policy in the UK over the last hundred years shows that public policy interventions can be remarkably effective.[2] The Forestry Commission was largely responsible for doubling the area of English woodland over the last hundred years from around 5 to 10 per cent. This was achieved by afforesting large tracts of land and encouraging the management and establishment of many private woodlands through tax concessions and grants. But people took a lot of persuading to accept the new areas of woodland. They were largely convinced by the desperate need for timber following the two world wars and the benefits for rural employment. Many, however, were horrified by the impact of the new plantations on the appearance of upland landscapes, on the freedom to walk across open fells and the damage caused to semi-natural heathlands and moors. The impact of future forests established for carbon sequestration will require robust defence of the likely benefits to ensure public backing for schemes. Extensive new areas of woodland may be acceptable in the abstract, but when linked to particular places with existing cultural values they are often problematic.

In broad terms there are six main types of interlinked woody landscapes. Let's consider some possible futures for these types, starting with the oldest and moving through to more recent ones. Ancient woodlands, as we saw in Chapter One, can be difficult to define but some individual woods have historical records going back several centuries. In the UK those with a proven history of being in existence for around four hundred years are defined as ancient, semi-natural woods. Some of these woods may never have been cleared and converted to another use such as agriculture and have a ground flora that links back to woodland plants which arrived in Britain after the last Ice Age. All have been managed for many centuries. Their value is now recognized by conservationists and foresters in many European countries, and such ancient woodlands should be protected from clearance and inappropriate management.

Second, we have recent plantations, which have been planted during the last few hundred years. The distinction between ancient and recent woodland has been complicated by the introduction in the UK of the category 'long-established woodland', meaning woodland in existence since 1893, which is currently being mapped (2024). Again, it is likely that recent plantations will remain as woodland, especially as most European countries have strict forest laws protecting the forest area. In some areas, however, where plantations have damaged existing semi-natural landscapes, such as some heaths, moors and upland pastures, management may need to be modified to increase conservation values. Where are large new areas of woodland to be planted? The expansion of European woodland in the last century was mainly in the mountains and uplands on moors and abandoned pastures, and on lowland heaths and the poor soils of karst landscapes. Such areas are no longer particularly important for agriculture but are increasingly popular for their landscapes and biodiversity. Crucially, if the area of woodland is to be increased significantly then many private landowners and farmers will have to be persuaded somehow to convert their valuable agricultural land to woodland.

Third is recent semi-natural woodland. These are woods that have not been planted, but which have grown up through natural regeneration. There are extensive areas of this type of woodland throughout many parts of Europe following rural depopulation and land abandonment. It is likely that the expansion of this type of woodland will continue, especially through the increasing popularity of diverse planned rewilding schemes. Forestry England announced in 2024, for example, that 8,000 hectares of their forest land are to be 'left to nature' to help improve biodiversity (illus. 27). There is no doubt that these schemes can have many benefits for nature conservation, such as the successful breeding of storks on the rewilded Knepp estate in Sussex.[3] In some areas this type of woodland should be discouraged, especially where it may damage the conservation value of existing semi-natural habitats such as chalk grasslands, hay meadows and heathland. This is also true of areas where the spread of such woodland increases the risk of damaging fires.

The fourth type of woody landscape is farmland trees. These include the wide range of trees growing within and alongside agricultural fields such as shelter belts, orchards, hedgerow trees, trees growing along rivers and streams, and small farm woods. While many of these trees were once an important part of local economies, to produce fuel, construction timber and many wood products, this link has often been broken. Traditional mixed combinations of trees and crops, for example *coltura promiscua* in Italy, where fruit trees and species such as elm and mulberry support growing vines, and the juxtaposition of large orchards of apple and pear trees with hay meadows and pasture in England, have become very scarce. Many farmland trees have been cleared away to allow modern forms of agriculture to be more cost-effective and such trees remain threatened.

Currently there is a great deal of support for an increase of trees on farmland, but it remains to be seen how farmers can be encouraged to allow this to happen. In Wales a government proposal that all farmers wishing to receive public funding under the new Sustainable Farming Scheme from 2025 should allow for 10 per cent of their farmland to be some form of woodland engendered a storm of protest from farmers who did not wish to lose productive farmland.[4] More positively it is undoubtedly true that the widening of existing hedgerows is one of the easiest and quickest ways of increasing tree cover on farmland. If carefully planned, considerable use could be made of natural regeneration of oaks, whose acorns are planted by jays and mice, or suckering species such as blackthorn, wild cherry and aspen. Local schemes such as the Fiesta del Chopo Cabecero, a festival to encourage the repollarding of ancient black poplars, held in Aguilar del Alfambra, Aragon, Spain, can be an effective way to raise the cultural appeal of tree management practices on the point of extinction.[5]

Garden and park trees form the fifth main type of woody landscape. These are extraordinarily diverse including cottage gardens with a few fruit trees, suburban gardens with ornamental trees and large private gardens and public parks. Most towns and cities have some eighteenth- and nineteenth-century houses with fine collections of ornamental trees. They also have very extensive and diverse twentieth- and twenty-first-century public and private suburban housing estates. In private gardens individual trees may have great sentimental value, being planted and nurtured by former family members. Trees are an essential attribute of public parks and gardens irrespective of their origin. Some, such as St James's Park and Hyde Park in London, derive from medieval hunting parks. Others, such as Kew Gardens, became centres of research and experimentation on the acclimatization of introduced trees. Bedgebury Pinetum (1925) and Derby Arboretum (1840) were established as labelled tree collections and other designed parks such as Maksimir Park, Zagreb (1794) and Regent's Park in London (1836) have excellent collections of trees. In the twentieth century Le Corbusier's idea of towers in the park became a dominant trope on both sides of the Iron Curtain and many thousands of trees were planted to enhance communal lawns.[6] Gardens and parks are places where trees are most likely to be closely observed, managed and pruned. They contain vast numbers of different species from around the world and can sometimes provide a haven for trees threatened with extinction. Some species such the ginkgo (*Ginkgo biloba*), dawn redwood (*Metasequoia glyptostroboides*) and the Wollemi pine (*Wollemia nobilis*), which was discovered in 1994 in New South Wales, are now more common in gardens and parks than in the wild.[7]

The sixth and final group of tree and woodland landscapes consists of urban and industrial trees. This is a very diverse category and includes many trees that we are most likely to see on a daily or weekly basis along roads, motorways and railways, and in retail parks and industrial estates, schools, sports grounds and cemeteries. These trees are often near at hand and familiar, but they are also mysterious. Who owns them? How are they managed? Some, such as street trees, are usually looked after by local councils, but others, including naturally regenerating trees on brownfield sites such as derelict factories and power stations, abandoned railway sidings and old gravel pits, remain uncared for. Such areas are often termed as 'scrub', which is a dangerous term as it demotes their value and gives people carte blanche to clear them for building. In fact, they have great potential for recreation and nature conservation and as a way of increasing tree cover. In Bologna, for example, natural regeneration of woodland at Prati di Caprara, on an old military area abandoned in the 1970s, is now much valued by local residents who are resisting its development.[8]

Many questions remain about the future of European woods. It is likely that native or long-established species that are known to grow well in the areas concerned will be favoured. The medium- to long-term impact of new tree diseases is a cause of much concern. The devastating upsurge in the number of tree diseases has been the result of lazily or weakly implemented import controls. Every attempt should be made to reduce the movement of infected young trees. Natural regeneration, which by definition uses local tree seeds, should be preferred over planting, which is expensive and often requires the use of plastic tree shelters. It may well be that species often thought of as invasive and problematic, such as goat willow (*Salix caprea*), sycamore (*Acer pseudoplatanus*), Norway maple (*Acer platanoides*), tree of heaven (*Ailanthus altissima*) and false acacia (*Robinia pseudoplatanus*), should be encouraged because of their rapid powers of natural regeneration.

Much research needs to be done to identify the best tree species to maximize the absorption of carbon. It may be tempting to make use of sequoias, eucalyptus or Sitka spruce, but the potential adverse impacts of these species will need to be carefully monitored. Much more thought needs to be given to the use of timber and wood products to increase the long-term carbon sequestration potential of woodlands. Charles Lamb, while writing a letter at his London desk during the Whitsun holiday in May 1819, remembered sadly that in previous years he would have been enjoying walking in the woods with his friends:

> What a reflection! Twelve years ago, and I should have
> kept that and the following holiday in the fields a-Maying.
> All of those pretty pastoral delights are over. This dead,
> everlasting dead desk – how it weighs the spirit of a
> gentleman down! This dead wood of the desk instead
> of your living trees![9]

But rather than contrast the dead wood with the living trees, we need to remember to connect the two: forests and the wood they produce are both required for life.

REFERENCES

INTRODUCTION

1 Lord Byron, *Childe Harold's Pilgrimage* (London, 1812), Canto II, pp. xxv–xxvi.
2 Ronald Blythe, *Next to Nature: A Lifetime in the English Countryside* (London, 2022), p. 425.
3 William Plomer, ed., *Kilvert's Diary* (London, 1944), p. 196.
4 Katherine Lack, *Herefordshire Farming through Time: Fellers, Tillers and Cider Makers* (Leominster, 2012), p. 197.
5 Forestry Commission, *Forestry Facts and Figures 2023* (Cheadle Heath, 2023).
6 R. W. Matthews et al., *Quantifying the Sustainable Forestry Carbon Cycle: Summary Report* (Farnham, 2022); H. Keith et al., 'Carbon Carrying Capacity in Primary Forests Shows Potential for Mitigation Achieving the European Green Deal 2030 Target', *Communications Earth and Environment*, V (2024), p. 256.
7 Andrew Fox, *Trees in Ancient Rome: Growing an Empire in the Late Republic and Early Principate* (London, 2023).
8 Della Hooke, *Trees in Anglo-Saxon England: Literature, Lore and Landscape* (Woodbridge, 2010); Charles Watkins, *Trees, Woods and Forests* (London, 2014); Fiona Stafford, *The Long, Long Life of Trees* (London, 2017).
9 Oliver Rackham, *Ancient Woodland* (London, 1980); G. F. Peterken and E. Mountford, *Woodland Development: A Long-Term Study of Lady Park Wood* (Wallingford, 2018); Keith Kirby and Charles Watkins, eds, *Europe's Changing Woods and Forests* (Wallingford, 2015); Keith Kirby and Jeanette Hall, *Woodland Survey Handbook* (London, 2019).
10 Keith Kirby and Charles Watkins, 'The Forest Landscape before Farming', in *Europe's Changing Woods and Forests*, ed. Kirby and Watkins, pp. 33–45; A. P. Molnár et al., 'Is There a Massive Glacial–Holocene Flora Continuity in Central Europe?', *Biological Reviews*, XCVIII (2023), pp. 2307–19.
11 N. Roberts et al., 'Europe's Lost Forests: A Pollen-Based Synthesis for the Last 11,000 Years', *Scientific Reports*, VIII (2018), art. 716.

12 S. Czerwiński et al., 'Synthesis of Palaeoecological Data from the Polish Lowlands Suggests Heterogeneous Patterns of Old-Growth Forest Loss after the Migration Period', *Scientific Reports*, XII (2022), p. 8559.

13 L. Östlund et al., 'Intensive Land Use in the Swedish Mountains between AD 800 and 1200 Led to Deforestation and Ecosystem Transformation with Long-Lasting Effects', *Ambio*, XLIV (2015), pp. 508–20.

1 ANCIENT ORIGINS

1 Gina Douglas, 'Linnaeus, Carl [later Carl von Linné] (1707–1778)', *Oxford Dictionary of National Biography* (Oxford, 2007).

2 Marianne North, *A Vision of Eden* (Exeter, 1980), pp. 87–8.

3 Floyd J. Franco, 'Native American Views and Values of Giant Sequoia', *Symposium on Giant Sequoias: Their Place in the Ecosystem and Society, June 23–5* (Visalia, CA, 1992), p. 64.

4 Gary D. Lowe, *Debunking the Sequoia Honoring Sequoyah Myth* (Livermore, CA, 2018); Nancy E. Muleady-Mecham, 'Endlicher and Sequoia: Determination of the Etymological Origin of the Taxon, *Sequoia*', *Bulletin of the Southern California Academy of Sciences*, CXVI (2017), pp. 137–46.

5 Ross Holland et al., 'Giant Sequoia (*Sequoiadendron giganteum*) in the UK: Carbon Storage Potential and Growth Rates', *Royal Society Open Science*, XI (2024).

6 Gideon F. Smith and Estrela Figueiredo, '"Rhodes-" Must Fall: Some of the Consequences of Colonialism for Botany and Plant Nomenclature', *Taxon*, LXXI (2022), pp. 1–5.

7 Shane Donald Wright and Len Norman Gillman, 'Replacing Current Nomenclature with Pre-Existing Indigenous Names in Algae, Fungi and Plants', *Taxon*, LXXI (2022), pp. 6–10.

8 James Main, *The Forest Planter and Pruner's Assistant* (London, 1839), pp. 6–7, 14; J. West, *Remarks on the Management or Rather the Mismanagement of Woods, Plantations and Hedgerow Timber* (Newark, 1842), p. 4.

9 John Standish and Charles Noble, *Practical Hints on Planting Ornamental Trees with Particular Reference to Coniferae* (London, 1852), p. 14.

10 J. Lindsay, 'Charcoal Iron Smelting and Its Fuel Supply: The Example of the Lorn Furnace, Argyllshire, 1733–1876', *Journal of Historical Geography*, I (1975), pp. 283–98.

11 E.J.T. Collins, 'Agriculture and Conservation in England; An Historical Overview', *Journal of the Royal Agricultural Society of England*, CXLVI (1985), pp. 38–46.

12 John Evelyn Denison, 'On the Grubbing Up of Woods', *Journal of the Royal Agricultural Society*, XVI (1855), pp. 352–528 (pp. 352, 323, 358).

13 William Schlich, *Forestry in the United Kingdom* (London, 1904), p. 61.

14 John Beckett and Charles Watkins, 'Natural History and Local History in Late Victorian and Edwardian England: The Contribution of the Victoria County History', *Rural History*, XXII (2011), pp. 59–87.

15 G. C. Druce, 'Botany', in *The Victoria History of the County of Northampton*, vol. I, ed. W. Ryland, D. Adkins and R. M. Serjeantson (London, 1902), p. 50;

Augustine Ley, 'Botany', in *The Victoria History of the County of Hereford*, vol. I, ed. William Page (London, 1908), p. 40.

16 Clement Reid, *The Origin of the British Flora* (London, 1899), pp. 15–17.

17 J. Hope-Simpson and D. Evans, 'Tansley, Sir Arthur George (1871–1955)', *Oxford Dictionary of National Biography* (Oxford, 2004).

18 C. E. Moss, W. M. Rankin and A. G. Tansley, 'The Woodlands of England', *New Phytologist*, IX (1910), pp. 113–49 (pp. 114–15).

19 A. G. Tansley, *The British Islands and Their Vegetation* (Cambridge, 1939), pp. 233–4.

20 B.R.G. Hammond, 'The Suppression of Coppice by Weeding', *Journal of the Forestry Commission*, XX (1949), pp. 112–13.

21 R. Carnell, 'Treatment of Felled Broadleaved Areas in the Midlands', *Journal of the Forestry Commission*, XXIV (1955), pp. 67–8, available at https://cdn. forestresearch.gov.uk.

22 H. M. Steven and A. Carlisle, *The Native Pinewoods of Scotland* (Edinburgh, 1959), p. v.

23 Oliver Rackham, *Hayley Wood, Its History and Ecology* (Cambridge, 1975); Oliver Rackham, *Trees and Woodland in the British Landscape* (London, 1976); Joyce Mildred Lambert, *The Making of the Broads: A Reconsideration of Their Origin in the Light of New Evidence* (London, 1960); E. Pollard, M. D. Hooper and N. W. Moore, *Hedges* (London, 1974); Colin Tubbs, *The New Forest: An Ecological History* (Newton Abbot, 1968); Ruth M. Tittensor, 'History of the Loch Lomond Oakwoods', *Scottish Forestry*, XXIV (1970), pp. 100–118; G. F. Peterken, 'Developmental Factors in the Management of British Woodlands', *Quarterly Journal of Forestry*, LXVIII (1974), pp. 141–9; Oliver Rackham, *Ancient Woodland: Its History, Vegetation and Use in England* (London, 1980), p. vii.

24 Ministry of Agriculture, Fisheries and Food, *The Reclamation of Derelict Woodland for Agricultural Use* (London, 1957), pp. iv, 19–20.

25 Charles Watkins, *Woodland Management and Conservation* (Newton Abbot, 1990), p. 6; Natural England, Ancient Woodland (England), see https:// naturalengland-defra.opendata.arcgis.com/datasets, accessed 3 June 2024.

26 Department for Environment, Food and Rural Affairs, Forestry Commission and Natural England, 'Keepers of Time: Ancient and Native Woodland and Trees Policy in England', Policy Paper, 27 May 2022.

27 Monika Wulf, 'Ancient Forests in Germany: Distribution, Importance for Maintaining Biodiversity, Protection and Threats', in *Ancient Woods, Trees and Forests: Ecology, History and Management*, ed. Alper H. Çolak, Simay Kırca and Ian D. Rotherham (London, 2023), pp. 124–44 (p. 139).

28 M. Biró et al., 'Behind the General Pattern of Forest Loss and Gain: A Long-Term Assessment of Semi-Natural and Secondary Forest Cover Change at Country Level', *Landscape and Urban Planning*, CCXX (2022), 104334.

29 Simay Kırca, Alper H. Çolak and Ian D. Rotherham, 'The Ancient Woodland Concept as a Practical Conservation Tool: The Turkish Experience', in *Ancient Woods, Trees and Forests: Ecology, History and Management*, ed. Çolak, Kırca and Rotherham, pp. 273–310.

30 S. Mollier et al., 'Historical Landscape Matters for Threatened Species in Temperate French Mountain Forests', *Biological Conservation*, CCLXIX (2022), 109544.

31 John Evelyn, *Sylva*, 2nd edn (London, 1670), p. 140.

32 Peter A. Thomas, 'Biological Flora of the British Isles: *Sorbus torminalis*', *Journal of Ecology*, CV (2017), pp. 1806–31; John Claudius Loudon, *Arboretum et fruticetum Britannicum* (London, 1854), vol. II, pp. 913–14; see, for example, Capovilla Distillati, Vicenza, 'Distillati di frutta rara e spontanea', www.capovilladistillati.it, accessed 3 June 2024.

33 Charles Watkins and Ben Cowell, *Letters of Uvedale Price*, Walpole Society, LXVII (2006), p. 190.

34 H. J. Elwes and Augustine Henry, *The Trees of Great Britain and Ireland*, vol. I (Edinburgh, 1906), pp. 153–4.

35 Tansley, *The British Islands and Their Vegetation*, p. 258; W. J. Bean, *Trees and Shrubs Hardy in the British Isles*, vol. III (London, 1951), p. 339; L. E. Whitehead, *Plants of Herefordshire: A Handlist* (Hereford, 1976).

36 Richard Mabey, *Flora Britannica* (London, 1996), p. 204; Roger Maskew, *Flora of Worcestershire* (Tenbury Wells, 2014), p. 221.

37 P. Pyttel, J. Kunz and J. Bauhus, 'Growth Regeneration and Shade Tolerance of the Wild Service Tree (*Sorbus torminalis* (L.) Crantz) in Aged Oak Coppice Forests', *Trees*, XXVII (2013), pp. 1609–19 (p. 1617).

38 Botanical Society of Britain and Ireland, 'BSBI Species Accounts Archive: *Sorbus Torminalis*', https://sppaccounts.bsbi.org, accessed 14 May 2024.

39 Christopher Guest, 'Wild Service Tree: Silviculture for High-Quality Timber Production', *Quarterly Journal of Forestry*, CXVII (2023), pp. 55–61.

40 K. K. Rasmussen and J. Kollman, 'Poor Sexual Reproduction on the Distribution Limit of the Rare Tree *Sorbus torminalis*', *Acta Oecologica*, XXV (2004), pp. 211–18.

41 Margaret Grainger, ed., *The Natural History Prose Writings of John Clare* (Oxford, 1983), p. 233; John Aubrey and Richard Rawlinson, *The Natural History and Antiquities of the County of Surrey: Begun in the Year 1673*, vol. III (London, 1718), p. 46.

42 Emily Sloan, *The Landscape Studies of Hayman Rooke (1723–1806): Antiquarianism, Archaeology and Natural History in the Eighteenth Century* (Woodbridge, 2018); Charles Watkins, *Trees, Woods and Forests* (London, 2014).

43 Hayman Rooke, *Descriptions and Sketches of Some Remarkable Oaks, in the Park at Welbeck, in the County of Nottingham, a Seat of His Grace the Duke of Portland* (London, 1790), pp. 5–6.

44 Evelyn, *Sylva*, p. 160.

45 Rooke, *Descriptions and Sketches*, pp. 16–17.

46 Kristof Haneca, Katarina Čufar and Hans Beeckman, 'Oaks, Tree-Rings and Wooden Cultural Heritage: A Review of the Main Characteristics and Applications of Oak Dendrochronology in Europe', *Journal of Archaeological Science*, XXXVI (2009), pp. 1–11; Charles Watkins, Christopher Lavers and Robert Howard, *Veteran Tree Management and Dendrochronology: Birklands and Bilhaugh CSAC, Nottinghamshire*, English Nature Research Reports 489 (Peterborough, 2003).

47 Rooke, *Descriptions and Sketches*, pp. 6–9.
48 Ibid., pp. 10–12.
49 Ibid., pp. 13–14.
50 Mat Collishaw, see https://matcollishaw.com, accessed 3 June 2024; Caroline Locke, see www.carolinelocke.org, accessed 3 June 2024.
51 Rooke, *Descriptions and Sketches*, p. 13.
52 T. Clayton, 'Miller, John Sebastian [formerly Johann Sebastian Müller] (1715–1792)', *Oxford Dictionary of National Biography* (Oxford, 2004).
53 John Evelyn, *Silva; or, A Discourse of Forest-Trees* (York, 1776). Quotation from 1801 edn, pp. 208–9.
54 J.M.W. Turner, *The Cowthorpe Oak from the South*, 1816, pencil on white wove paper, London, Tate Britain.
55 Jacob George Strutt, *Sylva Britannica; or, Portraits of Forest Trees Distinguished for Their Antiquity, Magnitude, or Beauty* (London, 1830), p. 35; Christiana Payne, 'Ancient Trees and Aged Peasants', *Interdisciplinary Studies in the Long Nineteenth Century*, XXXII (2021).
56 Strutt, *Sylva Britannica*, p. 26.
57 David Whitehead, 'Some Picturesque Influences upon the Study of Natural History in 19th-Century Herefordshire', *Transactions of the Woolhope Naturalists' Field Club*, LVII (2009), pp. 17–50; Susanne Seymour, Stephen Daniels and Charles Watkins, 'Estate and Empire: Sir George Cornewall's Management of Moccas, Herefordshire and La Taste, Grenada, 1771–1819', *Journal of Historical Geography*, XXIV (1998), pp. 313–51.
58 William Plomer, ed., *Kilvert's Diary, 1870–1879* (London, 1944), pp. 304–5.
59 Paul T. Harding and Tom Wall, eds, *Moccas, an English Deer Park* (Peterborough, 2000).
60 Buglife, 'Back from the Brink Species Summary: Moccas Beetle (*Hypebaeus flavipes*)', https://naturebftb.co.uk, accessed 3 June 2024; S. E. Tshernyshev, '*Hypebaeus cooteri* sp. nov., the Nemoral Species of Soft-Winged Flower Beetles (Coleoptera, Malachiidae) in North Asia', *Diversity*, XIV (2022), p. 875.
61 M. Twyman, 'Hullmandel, Charles Joseph (1789–1850)', *Oxford Dictionary of National Biography* (Oxford, 2004).

2 DIVERSE MODERNITY

1 John Clare, 'Firwood', in *John Clare: Poems Chiefly from Manuscript*, ed. Edmund Blunden and Alan Porter (London, 1920), p. 103.
2 Thomas De Quincey, 'Sketches of Life and Manners; from *The Autobiography of an English Opium-Eater*', *Tait's Edinburgh Magazine*, VII (1840), pp. 346–56 (p. 356). Quoted in Peter Dale and Brandon C. Yen, *Versed in Living Nature: Wordsworth's Trees* (London, 2022), pp. 110–11.
3 L. Östlund et al., 'The War on Deciduous Forest: Large-Scale Herbicide Treatment in the Swedish Boreal Forest, 1948 to 1984', *Ambio*, LI (2022), pp. 1352–6 (pp. 1352–3).
4 Ibid., pp. 1359, 1353, 1359.
5 Ibid., pp. 1359, 1360, 1363.

6 R. M. Brown, *Chemical Control of Weeds in the Forest* (London, 1975), p. 14.
7 P. Quelch and C. M. Mills, 'Planting Phases in Balgownie Wood, Fife: Evidence from Historic Woodland Survey', in *Plantations in Scotland*, ed. Coralie M. Mills (Perth, 2016), pp. 9–17.
8 Andrzej Szczepkowski, Łukasz Tyburski and Małgorzata Sułkowska, 'Monument Trees in the Kampinos National Park (Central Poland): A Review', *Folia Forestalia Polonica*, LXII (2020), pp. 210–19; Eivind Handegard et al., 'Identifying Old Norway Spruce and Scots Pine Trees by Morphological Traits and Site Characteristics', *Scandinavian Journal of Forest Research*, XXXVI (2021), pp. 550–62; V. Nolan et al., 'The Ancient Tree Inventory: A Summary of the Results of a 15 Year Citizen Science Project Recording Ancient, Veteran and Notable Trees across the UK', *Biodiversity and Conservation*, XXIX (2020), pp. 3103–29.
9 V. Ferreira et al., 'A Global Assessment of the Effects of Eucalyptus Plantations on Stream Ecosystem Functioning', *Ecosystems*, XXII (2019), pp. 629–42.
10 Olivier Vergnault, 'First UK Commercial Eucalyptus Plantation is in Cornwall', *Cornwall Live*, www.cornwalllive.com, 25 November 2019.
11 Paul Ames, 'Portugal's "Killer Forest": Deadly Wildfire Calls into Question Portugal's Embrace of Eucalyptus', *Politico*, www.politico.eu, 19 June 2017.
12 Andrew Long, *Aboriginal Scarred Trees in New South Wales: A Field Manual* (Hurstville, NSW, 2005).
13 F. J. Silva-Pando and R. Pino-Pérez, 'Introduction of *Eucalyptus* into Europe', *Australian Forestry*, LXXIX (2016), pp. 283–91.
14 John Claudius Loudon, *Arboretum et fruticetum Britannicum*, vol. II, 2nd edn (London, 1854), pp. 958–9.
15 H. J. Elwes and Augustine Henry, *The Trees of Great Britain and Ireland*, vol. IV (Edinburgh, 1906), pp. 1618, 1630 and 1640.
16 Silva-Pando and Pino-Pérez, 'Introduction of *Eucalyptus*', pp. 283–91.
17 Ignacio García-Pereda and Ana Duarte Rodrigues, 'Eucalyptus Acclimatisation for Fighting Malaria: Environmental and Medical Experiments in the Iberian Nineteenth Century', *Social History of Medicine*, XXXV (2021), pp. 72–96 (p. 73).
18 Deirdre Morris, 'Sir Ferdinand Jakob Heinrich von Mueller (1825–1896)', *Australian Dictionary of Biography* (Melbourne, 1974).
19 García-Pereda and Duarte Rodrigues, 'Eucalyptus Acclimatisation', p. 94. Mueller celebrated his friendship with Toverena, 'a nobleman who has given much encouragement to my researches', by naming a species of *Rhododendron* newly discovered in New Guinea after him: *The Victorian Naturalist: The Journal and Magazine of the Field Naturalist's Club of Victoria*, vol. I (1884), p. 101.
20 Ian Tyrrell, 'The Nature of G. P. Marsh: Tradition and Historical Judgement', *Environment and History*, X (2004), pp. 153–67 (p. 157).
21 One of the leading companies, Celbi, was established in 1965, see www.celbi.pt, accessed 3 June 2024.
22 Lars Kardell, Eliel Steen and António Fabião, 'Eucalyptus in Portugal – a Threat or a Promise?', *Ambio*, XV (1986), pp. 6–13 (pp. 11, 7 and 9).

23 Ibid., p. 12.

24 Ricardo J. Rodriguez, 'Portugal: 28 Years Ago a Village Fought against Eucalyptus. The Land Never Burned Again', *World Rainforest Movement Bulletin*, 238 (9 July 2018), www.wrm.org.uy.

25 García-Pereda and Duarte Rodrigues, 'Eucalyptus Acclimatisation', p. 72, fn. 3.

26 Ivan Tekić and Charles Watkins, 'Making Dalmatia Green Again: Reforestation at the "Horrible Edge" of Empire, 1870–1918', *Landscape History*, XLII (2021), pp. 99–118.

27 F. Ž. Kesterčanek, 'Prilozi za poviest šuma i šumskoga gospodarstva kod Hrvata', *Šumarski list*, VI (1882), pp. 117–35 (p. 121); J. Wessely, 'Kras hrvatske krajine i kako da se spasi, za tiem kraško pitanje uploške', *Šumarski list*, I (1877), pp. 57–109 (p. 59); J. Wessely, 'Kras hrvatske krajine i kako da se spasi, za tiem kraško pitanje uploške', *Šumarski list*, I (1877), pp. 229–52 (p. 244).

28 B. Kosović, 'Prvi šumarski stručni opis i nacrt šuma na Velebitu i Velikoj Kapeli od Dalmatinske medje do Mrkoplja i Ogulina', *Šumarski list*, XXXVIII (1914), pp. 4–16 (p. 13).

29 B. Anko, 'The Changing Role of Forest in the Karst Landscape', in *Human Influence on Forest Ecosystems Development in Europe*, ed. F. Salbitano (Bologna, 1989), pp. 95–109.

30 J. Wessely, 'Naše današnje znanje o brstu i brstiku', *Šumarski list*, III (1879), pp. 1–86.

31 HR-DASI-Šibenik 19.-20.st. 23 February 1904. *Dopis.* N. 1296; HR-DASI-Šumarstvo 19.-20.st. 22 August 1899. N. 80; HR-DASI-Šumarstvo 19.-20.st. 6 October 1907. *Iskaz pašnjaka šibenske općine koji bi se morali pošumiti u godini 1907/8 iznova.* N. 9931.

32 Elwes and Henry, *Trees*, vol. I (Edinburgh, 1906), p. 94; E. Baigent, 'Menzies, Archibald (1754–1842)', *Oxford Dictionary of National Biography* (Oxford, 2009); Loudon, *Arboretum*, vol. III, p. 2322.

33 Elwes and Henry, *Trees*, vol. I, p. 94.

34 A. C. Forbes, *English Estate Forestry* (London, 1904), p. 109.

35 Elwes and Henry, *Trees*, vol. I, pp. 94 and 96–7.

36 John D. Crozier, 'The Sitka Spruce as a Tree for Hill Planting and General Afforestation', *Transactions of the Royal Scottish Arboricultural Society*, XXIII (1910), pp. 7–16 (pp. 7, 9, 11, 16 and 13).

37 Ian Gambles, ed., *British Forests: The Forestry Commission, 1919–2019* (London, 2019), pp. 99–100.

38 Herbert L. Edlin, *Know Your Conifers* (London, 1970), p. 35.

39 Alan F. Mitchell, *Conifers in the British Isles: A Descriptive Handbook* (London, 1972), p. 178.

40 Herbert L. Edlin, *Forestry in Scotland* (Edinburgh, 1970), quoted in Ruth Tittensor, *Shades of Green: An Environmental and Cultural History of Sitka Spruce* (Oxford, 2016), p. 198.

41 Tittensor, *Shades of Green*, p. 220.

42 Ronald A. Knox, *Double Cross Purposes* (London, 1937), p. 12.

43 Ramblers' Association, *Afforestation: The Case against Expansion* (London, 1980), pp. 2, 3.

44 David A. Stroud et al., *Birds, Bogs and Forestry: The Peatlands of Caithness and Sutherland* (Peterborough, 1987), pp. 5–6; see also www.theflowcountry.org.uk, accessed 2 August 2024.

45 Forestry Commission, *68th Annual Report and Accounts for the Year Ended 31 March 1988* (London, 1989), p. 11.

46 Matthew Wilkinson, Georgios Xenakis and James Morison, 'The Carbon Balances of Two Contrasting Forest Stands Growing in the UK', *Forest Research*, Research Note 43 (2023), pp. 2–3 and 10–11.

47 Tittensor, *Shades of Green,* pp. 329 and 330.

48 'Climate Activists Pull Up Hundreds of Sitka Saplings on Coillte Land in North Leitrim', *Irish Independent*, 14 August 2023.

49 Nathalie Pettorelli, Sarah M. Durant and Johan T. du Toit, eds, *Rewilding* (Cambridge, 2019).

50 Frans Vera, *Grazing Ecology and Forest History* (Wallingford, 2000); Oliver Rackham, 'Savanna in Europe', in *The Ecological History of European Forests,* ed. Keith Kirby and Charles Watkins (Wallingford, 1998), pp. 1–24.

51 Frans Vera, 'Large-Scale Nature Development – the Oostvaardersplassen', *British Wildlife Supplement*, XX (2009), pp. 28–38 (p. 36).

52 J. Lorimer and C. Driessen, 'Wild Experiments at the Oostvaardersplassen: Rethinking Environmentalism in the Anthropocene', *Transactions of the Institute of British Geographers*, XXXIX (2014), pp. 169–81 (p. 179).

53 E. Bignal and D. McCracken, 'Herbivores in Space: Extensive Grazing Systems in Europe', *British Wildlife Supplement*, XX (2009), pp. 44–9 (p. 48).

54 National Trust, 'Wild Ennerdale Briefing Note', www.wildennerdale.co.uk, accessed 3 June 2024.

55 Rewilding Europe, 'Explore Our Rewilding Landscapes', www. rewildingeurope.com, accessed 3 June 2024.

56 Peter Parkes, 'A Pasture in Common: A Twentieth Century Environmental History of Ewyas Harold Common (Herefordshire)', *Rural History*, XVI (2005), pp. 111–32; Anthony Robinson, 'The Historical Ecology of Vegetation Change in Dovedale (Derbyshire and Staffordshire)', MRes, University of Nottingham, 2018.

57 Robert Hearn, Charles Watkins and Ross Balzaretti, 'The Cultural and Land Use Implications of the Reappearance of the Wild Boar in North West Italy: A Case Study of the Val di Vara', *Journal of Rural Studies*, XXXVI (2014), pp. 52–63.

3 ANIMALS AND TREES

1 Forestry England, 'Haugh Wood Forest Plan, 2020–2030', available at www.forestryengland.uk, accessed 3 June 2024.

2 Carolyn M. King and Roger A. Powell, *The Natural History of Weasels and Stoats* (Oxford, 2006), p. 120.

3 R. Robbins, 'Browne, Sir Thomas (1605–1682)', *Oxford Dictionary of National Biography* (Oxford, 2002); Sir Thomas Browne, *Pseudodoxia epidemica; or, Enquiries into Very Many Received Tenents and Commonly Presumed Truths* [1646], ed. Geoffrey Keynes (London, 1964), p. 201; Charles Snell,

'Status of the Common Tree Frog in Britain', *British Wildlife*, XVII (2006), pp. 153–60.

4 Alberto Gosa, 'Natural History Notes: Anura', *Herpetological Review*, XXXIV (2003), pp. 355–61.

5 S. O. Petrovan et al., 'Why Link Diverse Citizen Science Surveys? Widespread Arboreal Habits of a Terrestrial Amphibian Revealed by Mammalian Tree Surveys in Britain', *PLOS ONE*, XVII (2022), e0265156; C. Reading and G. Jofré, 'Habitat Selection and Range Size of Grass Snakes *Natrix natrix* in an Agricultural Landscape in Southern England', *Amphibia-Reptilia*, XXX (2009), pp. 379–88.

6 See 'Laboratory Confirmed Sightings of Yellow-Legged Hornet (Also Known as Asian Hornet) since 2016', www.gov.uk, accessed 25 July 2024.

7 Salvador Rebollo et al., 'Assessment of the Consumption of the Exotic Asian Hornet *Vespa velutina* by the European Honey Buzzard *Pernis apivorus* in Southwestern Europe', *Bird Study*, LXX (2023), pp. 136–50.

8 D. R. Risch, J. Ringma and M. R. Price, 'The Global Impact of Wild Pigs (*Sus scrofa*) on Terrestrial Biodiversity', *Scientific Reports*, XI (2021), 13256.

9 Homer, *Iliad*, trans. A. T. Murray and W. F. Wyatt (Cambridge and London, 1999), Book 9, pp. 433–5.

10 John Gay, *Fables* (London, 1727), ed. Vinton Dearing and Charles Beckwith (Oxford, 1974), pp. 307–8.

11 Robert Hearn, Charles Watkins and Ross Balzaretti, 'The Cultural and Land Use Implications of the Reappearance of the Wild Boar in North West Italy: A Case Study of the Val di Vara', *Journal of Rural Studies*, XXXVI (2014), pp. 52–63.

12 Ibid., p. 58.

13 Ibid.

14 Raquel Castillo-Contreras et al., 'Urban Wild Boars Prefer Fragmented Areas with Food Resources near Natural Corridors', *Science of the Total Environment*, DCXV (2018), pp. 282–8.

15 Pietro Piana et al., 'Urban Rewilding: Human-Wildlife Relations in Genoa, NW Italy', *Cities*, CXLIV (2024), 104660.

16 Forestry England, 'Background, Population and Management of Boar in the Forest of Dean', www.forestryengland.uk, updated 21 November 2023; Kieran O'Mahony, 'Blurring Boundaries: Feral Rewilding, Biosecurity and Contested Wild Boar Belonging in England', *Conservation and Society*, XVIII (2020), pp. 114–25.

17 Umberto Albarella, 'Pig Husbandry and Pork Consumption in Medieval England', in *Food in Medieval England: Diet and Nutrition*, ed. C. M. Woolgar, D. Serjeantson and T. Waldron (Oxford, 2006), pp. 72–87.

18 H. C. Darby, *Domesday England* (Cambridge, 1977); Oliver Rackham, *Ancient Woodland* (London, 1980), pp. 119 and 124; D. Jørgensen, 'Pigs and Pollards – Medieval Insights for UK Wood Pasture Restoration', *Sustainability*, V (2013), pp. 387–99.

19 Julie Hamilton and Richard Thomas, 'Pannage, Pulses and Pigs: Isotopic and Zooarchaeological Evidence for Changing Pig Management Practices in Later Medieval England', *Medieval Archaeology*, LVI (2012), pp. 234–59.

20 Andrew Margetts, 'In the Time of Mast: Investigating the Medieval Swine Pannage of South-East England', *Agricultural History Review*, LXX (2022), pp. 241–68 (p. 247); William Gilpin, *Remarks on Forest Scenery, and Other Woodland Views*, vol. II (London, 1791), pp. 112–17.

21 Colin R. Tubbs, *The New Forest* (London, 1986) pp. 111–12; see Forestry England, 'Pannage, Pigs and Acorns!', www.forestryengland.uk, accessed 3 June 2024.

22 University of Nottingham Department of Manuscripts and Special Collections (PW 1/643).

23 P. Seddon, 'Cavendish, Henry, Second Duke of Newcastle upon Tyne (1630–1691)', *Oxford Dictionary of National Biography* (Oxford, 2004).

24 Thomas Clarkson, *The History of the Rise, Progress, and Accomplishment of the Abolition of the African Slave-Trade by the British Parliament* (London, 1808) p. 84.

25 Nottinghamshire Archives DD/SR/211/266.

26 Zsolt Molnár et al., *Pigs in the Forest and Marsh: Traditional Ecological Knowledge of Svinjars* (Vácrátót, 2023), pp. 85, 87, 29, 32 and 35; A. Trouwborst, M. Krofel and J.D.C. Linnell, 'Legal Implications of Range Expansions in a Terrestrial Carnivore: The Case of the Golden Jackal (*Canis aureus*) in Europe', *Biodiversity and Conservation*, XXIV (2015), pp. 2593–610.

27 Molnár, *Pigs in the Forest and Marsh*, pp. 88, 113, 123, 162, 153, 129 and 185.

28 Robert Hearn, 'Gains and Losses in the European Mammal Fauna', in *Europe's Changing Woods and Forests*, ed. Keith Kirby and Charles Watkins (Wallingford, 2015) pp. 193–207.

29 Jamie Lorimer, 'Worlding and Weirding with Beaver: A More-Than-Human Political Ecology of Ecosystem Engineering', *Transactions of the Institute of British Geographers*, L (2024), e12698.

30 M. Wróbel and Krysztofiak-Kaniewska, 'Long-Term Dynamics of and Potential Management Strategies for the Beaver (*Castor fiber*) Population in Poland', *European Zoological Journal*, LXXXVIII (2020), pp. 116–21 (p. 118).

31 Martin Gaywood, 'Reintroducing the Eurasian Beaver *Castor fiber* to Scotland', *Mammal Review*, XLVIII (2018), pp. 48–61 (p. 48).

32 Kevin Jones et al., 'Willow (*Salix* spp.) and Aspen (*Populus tremula*) Regrowth after Felling by the Eurasian Beaver (*Castor fiber*): Implications for Riparian Woodland Conservation in Scotland', *Aquatic Conservation: Marine and Freshwater Ecosystems*, XIX (2009), pp. 75–87.

33 See Wild Ennerdale, 'Beavers', www.wildennerdale.co.uk, accessed 3 June 2024.

34 R. E. Auster, S. W. Barr and R. E. Brazier, 'Improving Engagement in Managing Reintroduction Conflicts: Learning from Beaver Reintroduction', *Journal of Environmental Planning and Management*, LXIV (2020), pp. 1713–34 (p. 1714).

35 Roger E. Auster et al., 'Learning to Live with Reintroduced Species: Beaver Management Groups Are an Adaptive Process', *Restoration Ecology*, XXXI (2023), e13899.

36 K. H. Baker et al., 'The 10,000-Year Biocultural History of Fallow Deer and Its Implications for Conservation Policy', *Proceedings of the National Academy of Sciences*, CXXI (2024), pp. 1–8 (p. 2).

37 P. Dolman et al., 'Escalating Ecological Impacts of Deer in Lowland Woodland', *British Wildlife*, XXI (2010), pp. 242–54 (p. 244).

38 Markus P. Eichhorn et al., 'Effects of Deer on Woodland Structure Revealed through Terrestrial Laser Scanning', *Journal of Applied Ecology*, LIV (2017), pp. 1615–26 (p. 1615).

39 Matthew Carter, 'The "Pheasant of the Future": Encountering and Imagining Reeves' Pheasant in Britain, 1831–1913', *Rural History*, XXXV (2024), pp. 277–93; F. Fan, 'Reeves, John (1774–1856)', *Oxford Dictionary of National Biography* (Oxford, 2010).

40 N. Chapman, S. Harris and A. Stanford, 'Reeves' Muntjac *Muntiacus reevesi* in Britain: Their History, Spread, Habitat Selection, and the Role of Human Intervention in Accelerating Their Dispersal', *Mammal Review*, XXIV (1994), pp. 113–60 (p. 120); M. S. Freemanet al., 'Population Genetics of Invasive Muntjac Deer', *Journal of Zoology*, CCXCVIII (2016), pp. 54–63.

41 G. Thomson and A. Baker, 'Russell, Herbrand Arthur, Eleventh Duke of Bedford (1858–1940)', *Oxford Dictionary of National Biography* (Oxford, 2014); K. Jordan and V. Quirke, 'Rothschild, Lionel Walter, Second Baron Rothschild (1868–1937)', *Oxford Dictionary of National Biography* (Oxford, 2004).

42 Chapman, Harris and Stafford, 'Reeves' Muntjac', pp. 121, 122, 124, 128 and 116.

43 A. S. Cooke and L. Farrell, 'Impact of Muntjac Deer (*Muntiacus reevesi*) at Monks Wood National Nature Reserve, Cambridgeshire, Eastern England', *Forestry*, LXXIV (2001), pp. 241–50.

44 See 'Reeves' muntjac deer' at www.mammal.org.uk.

45 Ibid.

46 A. I. Ward, S. Richardson and J. Mergeay, 'Reeves' Muntjac Populations Continue to Grow and Spread across Great Britain and Are Invading Continental Europe', *European Journal of Wildlife Research*, LXVII (2021), art. 34, p. 4, p. 1.

4 TREES UNDER ATTACK

1 S. S. Sayedi, B. W. Abbott and B. Vannière, 'Assessing Changes in Global Fire Regimes', *Fire Ecology*, XX (2024), art. 18, p. 1.

2 See Forest Research, 'Tools and Resources: Pest and Disease Resources', www.forestresearch.gov.uk, accessed 3 June 2024.

3 See 'Dothistroma Needle Blight (*Dothistroma septosporum*)', ibid.

4 Clive Brasier et al., 'Phytophthora: An Ancient, Historic, Biologically and Structurally Cohesive and Evolutionarily Successful Generic Concept in Need of Preservation', *IMA Fungus*, XIII (2022), art. 12.

5 See Forest Research, 'Tools and Resources: Ramorum Disease (*Phytophthora ramorum*)', www.forestresearch.gov.uk, accessed 3 June 2024; Joan Webber, '*Phytophthora ramorum* – a Developing Story', *Quarterly Journal of Forestry*, CXVI (2022), pp. 123–7.

6 See Forest Research, 'Tools and Resources: Phytophthora Disease of Alder (*Phytophthora alni*)', www.forestresearch.gov.uk, accessed 24 January 2025.

7 A. Pérez-Sierra et al., '*Phytophthora pluvialis*: A New Threat to Forestry?', *Quarterly Journal of Forestry*, CXVI (2022), pp. 128–30.

8   See Forest Research, 'Tools and Resources: Emerald Ash Borer Beetle
    (*Agrilus planipennis*)', www.forestresearch.gov.uk, accessed 3 June 2024;
    Mariella Marzano et al., 'Lessons from the Frontline: Exploring How
    Stakeholders May Respond to Emerald Ash Borer Management in
    Europe', *Forests*, XI (2020), art. 6.

9   James Lees-Milne, *Ancient as the Hills* (London, 1997), p. 77 [Friday,
    24 August 1973].

10 A. T. Gillanders, *Forest Entomology* (Edinburgh and London, 1908), pp. 93
    and 95; J. F. McDiarmid Clark, 'Eleanor Ormerod (1828–1901) as an
    Economic Entomologist: "Pioneer of Purity Even More Than of Paris
    Green"', *British Journal for the History of Science*, XXV (1992), pp. 431–52.

11 See Forest Research, 'Tools and Resources: Dutch Elm Disease (*Ophiostoma
    novo-ulmi*)', www.forestresearch.gov.uk, accessed 24 January 2025.

12 T. R. Peace, *The Status and Development of Elm Disease in Britain*
    (London, 1960), pp. 6–7.

13 *The Times*, 1 September 1928.

14 Ibid., 19 October 1932.

15 Ibid., 16 December 1935.

16 Ibid., 27 December 1938.

17 Ibid., 9 October 1943.

18 Ibid., 25 January 1955.

19 Peace, *Elm Disease*, p. 41.

20 J. N. Gibbs, D. A. Burdekin and C. Brasier, *Dutch Elm Disease* (London, 1977).

21 Clive Potter et al., 'Learning from History, Predicting the Future: The UK
    Dutch Elm Disease Outbreak in Relation to Contemporary Tree Disease
    Threats', *Philosophical Transactions of the Royal Society B*, CCCLXVI (2011),
    pp. 1966–74; Clive Potter, 'Tree and Forest Pests and Diseases: Learning
    from the Past', in *Europe's Changing Woods and Forests*, ed. Keith Kirby and
    Charles Watkins (Wallingford, 2015), pp. 337–46 (pp. 339–41).

22 A. J. Grayson, ed., *Broadleaves in Britain: Addresses, Supplementary Papers
    and Discussions* (Edinburgh, 1982), pp. 4–5, available at https://cdn.
    forestresearch.gov.uk.

23 Barnaby Wylder et al., 'Evidence from Mortality Dating of *Fraxinus
    excelsior* Indicates Ash Dieback (*Hymenoscyphus fraxineus*) Was Active
    in England in 2004–2005', *Forestry*, XCI (2018), pp. 434–43 (p. 434).

24 Barnaby Wylder, 'A Decade of Ash Dieback: Where Are We Now?',
    Quarterly Journal of Forestry, CXVII (2023), pp. 46–54 (pp. 47 and 52).

25 R. G. Pawsey, 'Ash Decline', in *Advances in Practical Arboriculture*,
    ed. D. Patch (London, 1987), pp. 156–9 (p. 156); S. K. Hull and J. N. Gibbs,
    Ash Dieback: A Survey of Non-Woodland Trees (London, 1991).

26 Wylder, 'Decade of Ash Dieback', pp. 47–8.

27 George Peterken et al., 'Ash Dieback in Woodland Nature Reserves',
    British Wildlife, XXXV (2024), pp. 431–7.

28 Forest Research, 'Tools and Resources: Ash Dieback (*Hymenoscyphus
    fraxineus*)', www.forestresearch.gov.uk, accessed 24 January 2025;
    James K. M. Brown and Elizabeth S. Orton, 'Ashes from Ashes', *Quarterly
    Journal of Forestry*, CXV (2021), pp. 107–13 (p. 113).

29 Forest Research, 'Tools and Resources: Sweet Chestnut Blight
    (*Cryphonectria parasitica*)', www.forestresearch.gov.uk, accessed 3 June 2024;
    D. Rigling and S. Prospero, '*Cryphonectria parasitica*, the Causal Agent of
    Chestnut Blight: Invasion History, Population Biology and Disease Control',
    Molecular Plant Pathology, XIX (2018), pp. 7–20.
30 Paolo Squatriti, *Landscape and Change in Early Medieval Italy: Chestnuts,
    Economy and Culture* (Cambridge, 2013).
31 Rigling and Prospero, '*Cryphonectria parasitica*', p. 11.
32 Forest Research, 'Sweet Chestnut Blight'.
33 Ibid.; Rigling and Prospero, '*Cryphonectria parasitica*', pp. 13, 8 and 14.
34 G. Gilioli et al., 'Modelling Local and Long-Distance Dispersal of Invasive
    Chestnut Gall Wasp in Europe', *Ecological Modelling*, CCLXII (2013), pp. 281–90.
35 E. Gehring et al., 'Chestnut Tree Damage Evolution Due to "*Dryocosmus
    kuriphilus*" Attacks', *Journal of Pest Science*, XCIII (2020), pp. 103–15; see Forest
    Research, 'Tools and Resources: Oriental Chestnut Gall Wasp (*Dryocosmus
    kuriphilus*)', www.forestresearch.gov.uk, accessed 3 June 2024.
36 S. S. Sayedi, B. W. Abbott and B. Vannière, 'Assessing Changes in Global Fire
    Regimes', *Fire Ecology*, XX (2024), art. 18, p. 1.
37 N. Fernandez-Anez et al., 'Current Wildland Fire Patterns and Challenges in
    Europe: A Synthesis of National Perspectives', *Air, Soil and Water Research*,
    XIV (2021), pp. 1 and 11.
38 Katarina Pavlek et al., 'Spatial Patterns and Drivers of Fire Occurrence in a
    Mediterranean Environment: A Case Study of Southern Croatia', *Geografisk
    Tidsskrift: Danish Journal of Geography*, CXVII (2017), pp. 22–35.
39 Diego Moreno, *Dal documento al terreno* (Bologna, 1990); Ross Balzaretti, Mark
    Pearce and Charles Watkins, eds, *Ligurian Landscapes* (London, 2004); Roberta
    Cevasco, *Memoria verde: Nuovi spazi per la geografia* (Reggio Emilia, 2007).
40 A. S. Mathews and F. Malfatti, 'Wildfires as Legacies of Agropastoral
    Abandonment: Gendered Litter Raking and Managed Burning as Historic
    Fire Prevention Practices in the Monte Pisano of Italy', *Ambio*, LIII (2024),
    pp. 1065–76 (p. 1068).
41 Ibid., p. 1071.
42 Ashleigh R. Harper et al., 'Prescribed Fire and Its Impacts on Ecosystem
    Services in the UK', *Science of the Total Environment*, DCXXIV (2018),
    pp. 691–703.
43 G. S. Robertson et al., 'Does Rotational Heather Burning Increase Red
    Grouse Abundance and Breeding Success on Moors in Northern England?',
    Wildlife Biology (2017), pp. 1–10.
44 Peter Sahlins, *Forest Rites: The War of the Demoiselles in Nineteenth-Century
    France* (Cambridge, MA, 1994), p. 2.
45 Jean-Paul Métailié, *Le Feu pastoral dans les Pyrénées centrales* (Toulouse, 1981).
46 P. M. Fernandes et al., 'Prescribed Burning in Southern Europe: Developing
    Fire Management in a Dynamic Landscape', *Frontiers in Ecology and the
    Environment*, XI (2013), e4–14 (e7).
47 Joachim Popek, 'Firewood and Timber: The Meaning of the Forest Common
    Rights in the Everyday Life of Peasants in Austrian Galicia', *Landscape
    History*, XLIV (2023), pp. 61–80 (p. 64).

48 Ibid., p. 65.

49 Ibid., pp. 67 and 69.

50 Natalia Ginzburg, *The Little Virtues* (London, 2018), p. 6; Valerio Varesi, *The Dark Valley* (London, 2012), p. 74.

5 FELLING TREES

1   R. Sikkema et al., 'A Market Inventory of Construction Wood for Residential Building in Europe – in the Light of the Green Deal and New Circular Economy Ambitions', *Sustainable Cities and Society*, CX (2023), 104370.

2   National Tree Safety Group, *Common Sense Risk Management of Trees: Guidance on Trees and Public Safety in the UK for Owners, Managers and Advisers* (Edinburgh, 2011).

3   Myles Hildyard, *It Is Bliss Here: Letters Home, 1939–1945* (London, 2005).

4   James Lees-Milne, *Caves of Ice* (London, 1983) p. 102 [28 October 1946].

5   Ibid., p. 159 [Sunday, 4 May 1947].

6   Alan Saville, ed., *Secret Comment: The Diaries of Gertrude Savile, 1721–1757* (Kingsbridge, 1997), p. 50 [Thursday, 10 August 1727].

7   William Wordsworth, *A Complete Guide to the Lakes* (Kendal and London, 1846), p. 89.

8   Anton Chekhov, *Uncle Vanya* [1899], trans. Constance Garnet (London, 1923).

9   H. Belsey, 'Gainsborough, Thomas (1727–1788)', *Oxford Dictionary of National Biography* (Oxford, 2004).

10 William Plomer, ed., *Kilvert's Diary, 1870–1879*, 3 vols (London, 1944), vol. III, p. 305 [Sunday, 23 April 1876]; vol. I, p. 135 [Tuesday, 18 July 1871]; vol. II, p. 131 [3 February 1872]; vol. II, p. 164 [Monday, 4 March 1872]; vol. II, pp. 168–9 [Tuesday, 26 March 1872].

11 Claude Colleer Abbott, ed., *The Correspondence of Gerard Manley Hopkins and Richard Watson Dixon* (Oxford, 1935), p. 26.

12 Robert Bernard Martin, *Gerard Manley Hopkins: A Very Private Life* (New York, 1991), pp. 305–7.

13 Ibid., pp. 11 and 56.

14 Humphry House, ed., *The Journals and Papers of Gerard Manley Hopkins* (London, 1959), p. 134 [4 and 5 May 1866]; p. 141 [29 June 1866].

15 Ibid., p. 192.

16 Ibid., p. 230 [8 April 1873].

17 Martin, *Hopkins*, p. 56.

18 Arnd Bohm, 'William Cowper's "The Poplar-Field" and Hopkins' "Binsey Poplars": The Politics of Pastoral', *Hopkins Quarterly*, XXX (2003), pp. 45–58.

19 Jude V. Nixon, 'Fathering Graces at Hampstead: Manley Hopkins' "The Old Trees" and Gerard Manley Hopkins' "Binsey Poplars"', *Victorian Poetry*, XLIV (2006), pp. 191–211 (pp. 194–5, p. 206).

20 Oliver Rackham, *Ancient Woodland* (London, 1980), p. 343.

21 John Claudius Loudon, *Arboretum et fruticetum Britannicum*, vol. III, 2nd edn (London, 1854), pp. 1645, 1647 and 1648.

22 See USDA Forest Service, 'Fishlake National Forest: Pando – (I Spread)', www.fs.usda.gov, accessed 3 June 2024.

23 Loudon, *Arboretum*, vol. III, pp. 1657–8.
24 William Pontey, *The Profitable Planter*, 4th edn (London, 1814), p. 218; Loudon, *Arboretum*, vol. III, p. 1658.
25 Pontey, *Profitable Planter*, pp. 75–6.
26 Loudon, *Arboretum*, vol. III, p. 1658.
27 H. J. Elwes and Augustine Henry, *The Trees of Great Britain and Ireland*, vol. VII (Edinburgh, 1906), p. 1822.
28 Martin, *Hopkins*, p. 307.
29 See J. Ewing Ritchie, *The Real Gladstone* (London, 1898), p. 200.
30 H. Matthew, 'Gladstone, William Ewart (1809–1898)', *Oxford Dictionary of National Biography* (Oxford, 2022).
31 Peter Sewter, 'W. E. Gladstone: A Love for Trees and Tree-Felling', MA thesis, University of Chester, 2007, fig. 1 facing p. 47; Peter Sewter, 'Gladstone as a Woodsman', in *William Gladstone: New Studies and Perspectives*, ed. R. Quinault, R. Swift and R. C. Windscheffel (London, 2013), pp. 155–76.
32 Sophieke Piebenga, 'William Sawrey Gilpin (1762–1843): Picturesque Improver', *Garden History*, XXII (1994), pp. 175–96.
33 Sewter, 'Gladstone: A Love for Trees', p. 71.
34 H.C.G. Matthew, ed., *The Gladstone Diaries*, 14 vols (London, 1968–94), vol. V, 29 October 1855, 31 July 1858, 2 August 1858, 5 August 1858, 24 August 1858; Roy Jenkins, *Gladstone* (London, 1995), p. 190.
35 Matthew, *Gladstone Diaries*, vol. VI, 23 December 1867, 25 December 1867, 27 December 1867, 28 December 1867, 2 December 1868.
36 Matthew, 'Gladstone, William Ewart'.
37 David M. Fahey, *The Politics of Drink in England, from Gladstone to Lloyd George* (Newcastle, 2022).
38 *Centennial Temperance Volume: A Memorial of the International Temperance Conference, Held in Philadelphia, June, 1876* (New York, 1877), p. 47; David Beckingham, *The Licensed City: Regulating Drink in Liverpool, 1830–1920* (Liverpool, 2017).
39 See Sewter, 'Gladstone: A Love for Trees', p. 12. Quoted in Philip Magnus, *Gladstone: A Biography* (New York, 1953), p. 193.
40 Quoted in Sewter, 'Gladstone: A Love for Trees', p. 22.
41 Alan Mitchell, *Trees of Britain and Northern Europe* (London, 1974), pp. 253–4; Elwes and Henry, *Trees*, vol. VII, pp. 1926–8.
42 Dr Richard Gaunt kindly provided this information.
43 Matthew, *Gladstone Diaries*, vol. IX, 10 May 1875, 11 May 1875.
44 Ibid., vol. IX, 22 October 1878, 3 November 1879, 4 April 1877; H. J. Leech, ed., *The Life of Mr Gladstone as Told by Himself* (London, 1893), p. 261, quoted in Sewter, 'Gladstone: A Love for Trees', p. 15.
45 Matthew, *Gladstone Diaries*, vol. X, 22 September 1881, 3 October 1881; vol. XI, 28 May 1889; vol. XII, 20 November 1889, 30 November 1889.
46 T.C.F., Kensington-Gardens, Letter to the Editor, *The Times*, 5 October 1880, p. 7.
47 W. H. Hudson, *Birds in London* (London, 1898), pp. 79–82.
48 Charlotte Mew, 'Men and Trees', *The Englishwoman*, XVII (1913), pp. 121–8.
49 Julia Copus, *This Rare Spirit: A Life of Charlotte Mew* (London, 2021), p. 330.

50 Charlotte Mew, 'The Trees Are Down', in *Collected Poems and Prose* (Manchester, 1981).

51 Copus, *Rare Spirit*, p. 347.

52 Thomas Hardy, *The Woodlanders* [1887] (London, 1969), p. 12.

53 Ibid., pp. 68–9.

54 Ibid., pp. 34 and 96–7.

55 Edward Thomas, *The Woodland Life* (London, 1897), p. 133.

56 Stan Smith, *Edward Thomas* (London, 1986), pp. 76–8.

57 Barbara A. Hanawalt, 'Peasant Women's Contribution to the Home Economy in Late Medieval England', in *Women and Work in Preindustrial Europe*, ed. Barbara A. Hanawalt (Bloomington, IN, 1986), pp. 7 and 9.

58 Mark Johnston, *The Tree Experts* (Oxford, 2021), p. 106.

59 William Alvis Brogden, *Ichnographia Rustica: Stephen Switzer and the Designed Landscape* (Oxford, 2017), pp. 2 and 45.

60 Johnston, *Tree Experts*, p. 148; Stephen Switzer, *Ichnographia Rustica; or, The Nobleman, Gentleman, and Gardener's Recreation*, vol. I (London, 1718), p. 73.

61 Sarah Law, Susanne Seymour and Charles Watkins, 'Women and Estate Management in the Early Eighteenth Century: Barbara Savile at Rufford Abbey, Nottinghamshire (1700–34)', *Rural History*, XXXIII (2022), pp. 23–39.

62 Barbara Savile, letter to George Savile, 3 March 1711, National Archives, Kew, NA DD/SR/211/435.

63 Ibid.

64 Thomas Smith, letter to Barbara Savile, National Archives, Kew, NA DD/SR/211/245, 20 December 1718.

65 Brian G. Awty, 'Denis Hayford', *Oxford Dictionary of National Biography* (Oxford, 2004); Brian G. Awty, 'Charcoal Ironmasters of Cheshire and Lancashire, 1600–1785', *Transactions of the Historic Society of Lancashire and Cheshire*, CIX (1957), pp. 71–124 (p. 84).

66 Barbara Savile, letter to George Savile, 3 March 1711.

67 Thomas Smith, letter to George Savile, 15 May 1721, National Archives, Kew, NA DD/SR/211/227/34; see also NA DD/SR/212/3.

68 Johnston, *Tree Experts*, p. 257.

69 List of members, *Transactions of the English Arboricultural Society* (1890), pp. 62–6.

70 Lars Östlund et al., 'Women in Forestry in the Early Twentieth Century: New Opportunities for Young Women to Work and Gain Their Freedom in a Traditional Agrarian Society', *Scandinavian Journal of Forest Research*, XXXV (2022), pp. 403–16 (pp. 403–4).

71 Helen E. Fitzrandolph and M. Doriel Hay, *The Rural Industries of England and Wales*, vol. II (Oxford, 1926), pp. 41 and 45–6.

72 Mavis Williams, *Lumber Jill* (Bradford on Avon, 1994), pp. 12 and 63.

73 Ibid., pp. 94–5, 70, 96 and 93.

74 Ibid., p. 92.

75 Forestry Commission, 'Gender Pay Gap Report: Report for the Year 2021–2022', see https://assets.publishing.service.gov.uk, March 2023; Forestry and Land Scotland, Blog, 'The Craft of Forestry Is Attracting a Growing Number of Women', https://forestryandland.gov.scot, 17 August 2022.

6 ARBOREAL AESTHETICS

1 Augustine Henry, *Forests, Woods and Trees in Relation to Hygiene* (London, 1919), pp. 14, 16–17, 18 and 23.
2 Philip Gardner, ed., *The Journals and Diaries of E. M. Forster*, vol. I (London, 2011), p. 136.
3 J. Pretty et al., 'Green Exercise in the UK Countryside: Effects on Health and Psychological Well-Being, and Implications for Policy and Planning', *Journal of Environmental Planning and Management*, L (2007), pp. 211–31; Dorit Karla Haubenhofer et al., 'The Development of Green Care in Western European Countries', *Explore*, VI (2010), pp. 106–11; Wendy Masterton et al., 'Greenspace Programmes for Mental Health: A Survey Study to Test What Works, for Whom, and in What Circumstances', *Health and Place*, LXXII (2021), 102669.
4 L. Bickerdike et al., 'Social Prescribing: Less Rhetoric and More Reality. A Systematic Review of the Evidence', *BMJ Open*, VII (2017), e013384.
5 Mea Allen, *E. A. Bowles and His Garden at Myddelton House, 1865–1954* (London, 1973), p. 77.
6 T. Marquina, R. K. Gould and D. Murdoch, '"Hey, tree. You are my friend": Assessing Multiple Values of Nature through Letters to Trees', *People and Nature*, V (2023), pp. 415–30.
7 Walt Whitman, *Specimen Days* [1887] (Oxford, 2023), p. 17.
8 Ibid., p. 94.
9 Ibid., pp. 95 and 99.
10 Ibid., pp. 103 and 128.
11 Ibid., pp. 104 and 184.
12 Ibid., pp. 121 and 175.
13 Howard Nelson, 'Timber Creek', in *Walt Whitman: An Encyclopedia*, ed. J. R. LeMaster and Donald D. Kummings (New York, 1998).
14 E. M. Forster, *Collected Short Stories* (London, 1947), pp. 5, 12–13, 33, 5 and 95.
15 Philip Gardner, ed., *The Journals and Diaries of E. M. Forster*, vol. I (London, 2011), p. 122.
16 Ibid., p. 136; E. M. Forster, *A Room with a View* [1908] (London, 2012), pp. 118, 134–5 and 114–15.
17 Forster, *Short Stories*, p. 63. See also Frederick Williams, 'Daphne Transformed: Parthenius, Ovid, and E. M. Forster', *Hermathena*, 166 (1999), pp. 45–62.
18 E. M. Forster, 'West Hackhurst: A Surrey Ramble', in *Journals and Diaries of E. M. Forster*, ed. Gardner, vol. III (London, 2011), p. 182.
19 E. M. Forster, 'My Wood, or the Effects of Property upon Character', *New Leader*, XIII (15 October 1926), p. 3. Later published as 'My Wood' in E. M. Forster, *Abinger Harvest* (London, 1936), pp. 33–6.
20 Forster, *Abinger Harvest*, p. 33.
21 Ibid., p. 34.
22 Ibid., p. 36.
23 'West Hackhurst: A Surrey Ramble', p. 183.
24 P. N. Furbank, *E. M. Forster: A Life*, vol. II (London, 1978), p. 198.
25 Forster, *Abinger Harvest*, p. 383.

26 Ibid., p. 395.
27 Ibid., p. 390.
28 Ibid., p. 399.
29 E. M. Forster, 'Havoc', in *Britain and the Beast*, ed. Clough Williams-Ellis (London, 1937), pp. 44–5.
30 D. Monk, 'E. M. Forster's Will: An Overlooked Posthumous Publication', *Legal Studies,* XXXIII (2013), pp. 572–97 (p. 582); Furbank, *Forster,* p. 199.
31 Gardner, *Journals and Diaries of E. M. Forster*, vol. III, p. 87 [Friday, 13 October 1950]; E. M. Forster, *Howard's End* [1910] (London, 2000), pp. 289–90.
32 J. L. García-Castaño, A. Terrab and M. A. Ortiz, 'Patterns of Phylogeography and Vicariance of *Chamaerops humilis L.* (Palmae)', *Turkish Journal of Botany,* XXXVIII (2014), pp. 1132–46; C. Chao and R. Krueger, 'The Date Palm (*Phoenix dactylifera L.*): Overview of Biology, Uses, and Cultivation', *HortScience,* XLII (2007), pp. 1077–82.
33 Marco Cassioli, 'La palma nel paesaggio agrario dell'estremo Ponente Ligure: il territorio di Sanremo alla fine del Medioevo', in *Palme di Liguria*, ed. Claudio Littardi (Rome, 2015), pp. 82–9.
34 Pietro Piana, Charles Watkins and Ross Balzaretti, 'Palm Landscapes', *Landscapes,* XIX (2018), pp. 43–65; Alessandro Carassale, 'Le palme di Sanremo e Bordighera: normative di raccolta e circuiti commerciali (secc. XV–XX)', in *Palme di Liguria*, ed. Littardi, pp. 93–150 (p. 101).
35 Davide Bertolotti, ed., *Viaggio nella Liguria Marittima*, vol. III (Turin, 1834), p. 277.
36 F. F. Hamilton, *Bordighera and the Western Riviera* (London, 1883), p. 331; Claudio Littardi, 'Il palmeto storico dell'estremo Ponente Ligure: evoluzione, forma e utilizzo', in *Palme di Liguria*, ed. Littardi, pp. 20–28 and 71.
37 Carassale, 'Le palme di Sanremo', p. 138; Littardi, 'Il palmetto storico', n. 126.
38 Giovanni Ruffini, *Doctor Antonio* (London, 1855), pp. 162 and 241.
39 Henry Alford, *The Riviera: Pen and Pencil Sketches from Cannes to Genoa* (London, 1870), p. 71.
40 Ibid., pp. 72–5.
41 A. Berger, *Alphabetical Catalogue of Plants Growing in the Garden of the Late Sir Thomas Hanbury at La Mortola, Ventimiglia, Italy* (London, 1912).
42 *Journal de Bordighera*, 8 December 1898; Luigi Viacava, *Lodovico Winter: giardiniere di Bordighera* (Genoa, 1996).
43 Robert Smail Jack and Fritz Scholz, ed. and trans., *Wilhelm Ostwald: The Autobiography* (Cham, 2017), pp. 260–61.
44 A. Gajone, *Nervi, Sant'Ilario e Quinto al Mare, Storia e Turismo* (Borgo S. Dalmazzo, 1955).
45 R. Thomson, *Monet and Architecture* (London, 2018), pp. 220 and 68; H. T. Day and H. Sturges, *Joslyn Art Museum: Paintings and Sculpture from the European and American Collections* (Omaha, NE, 1987), p. 100.
46 Guglielmi Libereso, *Libereso il giardiniere di Calvino* (Rome, 2013).
47 John Evelyn, *Sylva*, 2nd edn (London, 1670), p. 70.
48 William Parsons Diary, 14 March 1850, Nottingham University Library, Department of Manuscripts and Special Collections, MS 489/5; 21 November 1850, MS 489/6.

49 Paul A. Elliott, *British Urban Trees: A Social and Cultural History, c. 1800–1914* (Winwick, 2016), pp. 205 and 253; Jan Woudstra and Camilla Allen, eds, *The Politics of Street Trees* (London, 2022).

50 See www.fellingfilm.com, accessed 3 June 2024.

51 Melvyn Jones, *Sheffield's Woodland Heritage* (Sheffield, 2008); I. D. Rotherham and M. Flinders, 'No Stump City: The Contestation and Politics of Urban Street-Trees: A Case Study of Sheffield', *People, Place and Policy*, XII (2019), pp. 188–203.

52 Simon Crump, Calvin Payne and Julie Stribley, *Persons Unknown: The Battle for Sheffield's Street Trees* (Sheffield, 2022), p. 17.

53 Ibid., pp. 17 and 5.

54 Ibid., pp. 65–7.

55 Ibid., pp. 76–7, 128 and 155.

56 Ibid., p. 108.

57 See 'Sheffield Street Tree Partnership Strategy', May 2021, available at www.sheffield.gov.uk.

58 See 'Pupils of Western Road Council School (Trees)', 16 July 2020, available at www.warmemorialsonline.org.uk.

59 See Sheffield City Council, 'Decision Details: War Memorial Trees', 19 December 2017, available at https://democracy.sheffield.gov.uk.

60 Matthew Flinders and Matthew Wood, 'Ethnographic Insights into Competing Forms of Co-Production: A Case Study of the Politics of Street Trees in a Northern English City', *Social Policy and Administration*, LIII (2019), pp. 279–94.

7 SACRED TREES

1 Adam Stout, *The Glastonbury Thorn: Story of a Legend* (Glastonbury, 2020).

2 Vaughan Cornish, *Historic Thorn Trees in the British Isles* (London, 1941), pp. 9, 73 and 9.

3 John Claudius Loudon, *Arboretum et fruticetum Britannicum*, vol. II, 2nd edn (London, 1854), pp. 830, 833 and 844.

4 Stout, *Glastonbury Thorn*, p. 11, plate 16.

5 T. Sackett, 'Frith, Francis (1822–1898)', *Oxford Dictionary of National Biography* (Oxford, 2006).

6 H. J. Elwes and Augustine Henry, *The Trees of Great Britain and Ireland*, vol. III (Edinburgh, 1908), p. 454, fn. 2.

7 Ibid., p. 457.

8 E. W. Beals, 'The Remnant Cedar Forests of Lebanon', *Journal of Ecology*, LIII (1965), pp. 679–94 (pp. 680 and 691).

9 John Gerard, *The Herball or Generall Historie of Plantes* (London, 1597), p. 1113.

10 Loudon, *Arboretum*, vol. IV, p. 2409.

11 Ibid.

12 Henry Maundrell, *A Journey from Aleppo to Jerusalem: At Easter, AD 1697*, 7th edn (Oxford, 1749), p. 142; Robin Butlin, 'Maundrell, Henry (bap. 1665, d. 1701)', *Oxford Dictionary of National Biography* (Oxford, 2004).

13 Richard Pococke, *A Description of the East*, vol. II (London, 1745), part 1, p. 105.

14 John Evelyn, *Sylva*, 2nd edn (London, 1670), p. 116.

15 Ibid., pp. 120–21.

16 Ibid., p. 130.

17 Ibid., 3rd edn (London, 1679), p. 125.

18 Ibid., pp. 135 and 139.

19 Lara Hajar et al., '*Cedrus libani* (A. Rich) Distribution in Lebanon: Past, Present and Future', *Comptes Rendus Biologies*, CCCXXXIII (2010), pp. 622–30.

20 Marcello Rossi, 'Can Lebanon's Cedars Outlive Climate Change and a Pesky Insect?', *Al Jazeera News*, www.aljazeera.com, 20 April 2019.

21 Some identify the Turkish cedars of Lebanon as a subspecies: *Cedrus libani* ssp. *stenocoma*. See A. Farjon, *World Checklist and Bibliography of Conifers*, 2nd edn (London, 2001).

22 Melih Boydak, 'Regeneration of Lebanon Cedar (*Cedrus libani* A. Rich.) on Karstic Lands in Turkey', *Forest Ecology and Management*, CLXXVIII (2003), pp. 231–43; Nicklas Jansson, Ogün Ç. Türkay and Mustafa Avci, 'A Hidden Treasure in Turkey: Old Oaks of Unique Value', in *Ancient Woods, Trees and Forests*, ed. Alper H. Çolak, Simay Kirka and Ian D. Rotherham (London, 2023), pp. 335–44 (p. 339).

23 F. Nigel Hepper, 'The Cultivation of the Cedar of Lebanon in Western European Parks and Gardens from the 17th to the 19th Century', *Arboricultural Journal*, XXV (2001), pp. 197–219.

24 Loudon, *Arboretum*, vol. IV, pp. 2412–14.

25 G. Toomer, 'Pococke, Edward (1604–1691)', *Oxford Dictionary of National Biography* (Oxford, 2008); Andrew Lake, 'The First Protestants in the Arab World: The Contribution to Christian Mission of the English Aleppo Chaplains, 1597–1782', PhD thesis, University of Melbourne, 2015, p. 41.

26 Edward Pococke, *A Commentary on the Prophecy of Hosea* (Oxford, 1685), p. 807, quoted in Simon Mills, 'Edward Pococke (1604–1691), Comparative Arabic-Hebrew Philology, and the Bible', *Journal of Medieval and Early Modern Studies*, LIII (2023), pp. 130–31.

27 A. Murdoch, 'Campbell, Archibald, Third Duke of Argyll (1682–1761)', *Oxford Dictionary of National Biography* (Oxford, 2006); Michael Symes, Alison Hodges and John Harvey, 'The Plantings at Whitton', *Garden History*, XIV (1986), pp. 138–72.

28 Brilliana Harley, '"Functional Picturesque": Richard Payne Knight and Uvedale Price in Herefordshire', *Georgian Group Journal*, XXIV (2016), pp. 135–58; Charles Watkins and Ben Cowell, *Uvedale Price, 1747–1829: Decoding the Picturesque* (Woodbridge, 2012).

29 Margaret Stone et al., 'An Eighteenth-Century Obsession: The Plant Collection of the 6th Earl of Coventry at Croome Park, Worcestershire', *Garden History*, XLIII (2015), pp. 74–125; see 'Rotunda, Croome Park', National Trust Heritage Records Online, https://heritagerecords. nationaltrust.org.uk, accessed 3 June 2024.

30 William Gilpin, *Remarks on Forest Scenery and Other Woodland Views*, vol. I (London, 1791), pp. 73–6.

31 Ibid., pp. 77–8; T.F.T. Baker, J. S. Coburn and R. B. Pugh, ed., *A History of the County of Middlesex*, Victoria County History, vol. IV (London, 1971), p. 62; Loudon, *Arboretum*, vol. IV, p. 2417.

32 Jacob George Strutt, *Sylva Britannica; or, Portraits of Forest Trees Distinguished for Their Antiquity, Magnitude, or Beauty* (London, 1830), pp. 103, 106 and 108–9.

33 W. J. Bean, *Trees and Shrubs Hardy in the British Isles*, vol. I, 7th edn (London, 1950), p. 394.

34 Loudon, *Arboretum*, vol. IV, p. 2417.

35 Walter Nicol, *The Practical Planter* (London, 1803), pp. 50–51.

36 Elwes and Henry, *Trees*, vol. III, p. 467.

37 G. E. Mitton and J. C. Geikie, *Hammersmith, Fulham and Putney* (London, 1903), p. 49.

38 Loudon, *Arboretum*, vol. III, p. 2418.

39 William Ablett, *English Trees and Tree Planting* (London, 1880), p. 35; J. Blenkarn, *British Timber Trees* (London, 1859), p. 72.

40 Paul Elliott, Charles Watkins and Stephen Daniels, *The British Arboretum* (London, 2011), pp. 171–2.

41 John Claudius Loudon, *On the Laying Out, Planting, and Managing of Cemeteries; and on the Improvement of Churchyards* (London, 1843), p. 89.

42 P. Joyce, *A Guide to Abney Park Cemetery* (London, 1994).

43 G. Collison, *Cemetery Interment* (London, 1840), p. 297.

44 I. Rivers, 'Watts, Isaac (1674–1748)', *Oxford Dictionary of National Biography* (Oxford, 2008).

45 Collison, *Cemetery Interment*, pp. 305–7.

46 M. J. Campbell, *John Martin, Visionary Printmaker* (New York, 1992).

47 M. Myrone and A. Austen, 'Catalogue', in *John Martin: Apocalypse*, ed. M. Myrone (London, 2012), pp. 61–213.

48 Peter Dale and Brandon C. Yen, *Versed in Living Nature: Wordsworth's Trees* (London, 2022), pp. 72–3.

49 Ed Cumming, *John Wordsworth and the Wreck of the Earl of Abergavenny* (Portsmouth, 2016), available at www.nauticalarchaeologysociety.org.

50 Charles I. Patterson, 'The Meaning and Significance of Wordsworth's "Peele Castle"', *Journal of English and Germanic Philology*, LVI (1957), pp. 1–9 (fn. 3); William Wordsworth, letter to Sir George Beaumont, 1 August 1806, in *The Letters of William and Dorothy Wordsworth*, vol. II, ed. Ernest de Sélincourt, part 1, 2nd edn (Oxford, 2000).

51 Dale and Yen, *Versed in Living Nature*, p. 124.

52 Ibid., pp. 219–20.

53 Pietro Piana, Charles Watkins and Ross Balzaretti, *Rediscovering Lost Landscapes: Topographical Art in North-West Italy, 1800–1920* (Woodbridge, 2021).

54 Catherine Maria Fanshawe, letter to Catherine M. Bury, Countess of Charleville, Rome, 11 January 1820, Nottingham University Library, Department of Manuscripts and Special Collections, My 269.

55 F. Owen and D. B. Brown, *Collector of Genius: A Life of Sir George Beaumont* (New Haven, CT, and London, 1988); S. Avery-Quash and J. Sheldon, *Art for the Nation: The Eastlakes and the Victorian Art World* (London, 2011).

56 Owen and Brown, *Collector of Genius*, p. 204.
57 William Wordsworth, 'The Pine of Monte Mario at Rome, Text with Some Notes', in *The Poetical Works of William Wordsworth*, vol. VIII, ed. William Knight (London, 1896), pp. 58–9.
58 T. Sadler, ed., *Diary, Reminiscences, and Correspondence of Henry Crabb Robinson*, vol. III (London, 1869), pp. 116–24.
59 Ibid., pp. 116–17; *Poetical Works of William Wordsworth*, ed. Knight, vol. VIII, pp. 58–9; William Wordsworth, letter to Dorothy Wordsworth, 27 April 1837, in *Letters of William and Dorothy Wordsworth*, vol. VI, part 3 (Oxford, 1982).
60 William Wordsworth, letter to Mary and Dorothy Wordsworth, 6 May 1837, ibid.
61 *Poetical Works of William Wordsworth*, ed. Knight, vol. VIII, pp. 58–9.
62 Ernest de Sélincourt and Helen Darbishire, *The Poetical Works of William Wordsworth* (London, 1954), pp. 212–13.
63 *Poetical Works of William Wordsworth*, ed. Knight, vol. VIII, pp. 58–9.

CONCLUSION

1 George Shaw, *My Back to Nature* (London, 2016).
2 S. Raum and Clive Potter, 'Forestry Paradigms and Policy Change: The Evolution of Forestry Policy in Britain in Relation to the Ecosystem Approach', *Land Use Policy*, XLIX (2015), pp. 462–70.
3 See Forestry England, 'New Wild Areas for Nature Recovery in the Nation's Forests', www.forestryengland.uk, accessed 3 June 2024; Isabella Tree, *The Book of Wilding: A Practical Guide to Rewilding, Big and Small* (London, 2023).
4 See 'Sustainable Farming Scheme – Outline Proposals from 2025: Frequently Asked Questions', 29 September 2022, www.gov.wales.
5 See 'Chopos cabeceros', https://parquechopocabecero.com, accessed 3 June 2024.
6 Paul Elliott, Charles Watkins and Stephen Daniels, *The British Arboretum* (London, 2011); N. Tandarić, C. Watkins and C. D. Ives, 'Urban Planning in Socialist Croatia', *Hrvatski Geografski Glasnik*, LXXXI (2019), pp. 5–41.
7 Peter Crane, *Ginkgo: The Tree That Time Forgot* (New Haven, CT, 2013).
8 Andrea Zinzani and Enrico Curzi, 'Urban Regeneration, Forests and Socio-Environmental Conflicts: The Case of Prati di Caprara in Bologna, Italy', *ACME: An International Journal for Critical Geographies*, XIX (2020), pp. 163–86.
9 Charles Lamb, letter to Thomas Manning, 28 May 1819, in *The Works of Charles and Mary Lamb*, vols VI–VII (London, 1905), Letter 234, p. 522.

SELECT BIBLIOGRAPHY

Balzaretti, Ross, et al., eds, *Ligurian Landscapes: Studies in Archaeology, Geography and History* (London, 2004)

Bean, William J., *Trees and Shrubs Hardy in the British Isles*, 3 vols, 7th edn (London, 1951)

Cevasco, Roberta, *Memoria verde: nuovi spazi per la geografia* (Reggio Emilia, 2007)

Çolak, Alper H., et al., eds, *Ancient Woods, Trees and Forests: Ecology, History and Management* (London, 2023)

Collins, E.J.T., 'Agriculture and Conservation in England: An Historical Overview', *Journal of the Royal Agricultural Society of England*, CXLVI (1985), pp. 38–46

Crump, Simon, et al., *Persons Unknown: The Battle for Sheffield's Street Trees* (Sheffield, 2022)

Dale, Peter, and Brandon C. Yen, *Versed in Living Nature: Wordsworth's Trees* (London, 2022)

Edlin, Herbert, *Forestry and Woodland Life* (London, 1947)

Elliott, Paul A., *British Urban Trees: A Social and Cultural History, c. 1800–1914* (Winwick, 2016)

——, et al., *The British Arboretum: Trees, Science and Culture in the Nineteenth Century* (London, 2011)

Elwes, H. J., and Augustine Henry, *The Trees of Great Britain and Ireland*, 7 vols (Edinburgh, 1906–13)

Evelyn, John, *Sylva* (London, 1664, repr. 1670, 1706)

Forster, E. M., *Abinger Harvest* (London, 1936)

Fox, Andrew, *Trees in Ancient Rome: Growing an Empire in the Late Republic and Early Principate* (London, 2023)

Gilpin, William, *Remarks on Forest Scenery and Other Woodland Views* (London, 1791)

Harding, Paul T., and Tom Wall, eds, *Moccas: An English Deer Park* (Peterborough, 2000)

Hearn, Robert, et al., 'The Cultural and Land Use Implications of the Reappearance of the Wild Boar in North West Italy: A Case Study of the Val di Vara', *Journal of Rural Studies*, XXXVI (2014), pp. 52–63

Hooke, Della, *Trees in Anglo-Saxon England: Literature, Lore and Landscape* (Woodbridge, 2010)

Johnston, Mark, *The Tree Experts* (Oxford, 2021)

Kirby, Keith, and Charles Watkins, eds, *The Ecological History of European Forests* (Wallingford, 1998)

——, *Europe's Changing Woods and Forests* (Wallingford, 2015)

Law, Sarah, et al., 'Women and Estate Management in the Early Eighteenth Century: Barbara Savile at Rufford Abbey, Nottinghamshire (1700–34)', *Rural History*, XXXIII (2022), pp. 23–39

Loudon, John Claudius, *Arboretum et fruticetum Britannicum*, 8 vols (London, 1854)

Métaillé, J.-P., *Le Feu pastoral dans les Pyrénées centrales* (Toulouse, 1981)

Molnár, Zsolt, et al., *Pigs in the Forest and Marsh: Traditional Ecological Knowledge of Svinjars* (Vácrátót, 2023)

Moreno, Diego, *Dal documento al terreno* (Bologna, 1990)

Peterken, George, *Natural Woodland* (Cambridge, 1996)

——, *Trees and Woodlands* (London, 2023)

——, and Edward Mountford, *Woodland Development: A Long-Term Study of Lady Park Wood* (Wallingford, 2018)

Pettorelli, Nathalie, et al., eds, *Rewilding* (Cambridge, 2019)

Piana, Pietro, et al., *Rediscovering Lost Landscapes: Topographical Art in North-West Italy, 1800–1920* (Woodbridge, 2021)

Price, Uvedale, *Essays on the Picturesque* (London, 1810)

Rackham, Oliver, *Hayley Wood, Its History and Ecology* (Cambridge, 1975)

——, *Trees and Woodland in the British Landscape* (London, 1976)

——, *Ancient Woodland* (London, 1980)

Reid, Clement, *The Origin of the British Flora* (London, 1899)

Roberts, N., et al., 'Europe's Lost Forests: A Pollen-Based Synthesis for the Last 11,000 Years', *Scientific Reports*, VIII (2018), art. 716

Sahlins, Peter, *Forest Rites: The War of the Demoiselles in Nineteenth-Century France* (Cambridge, MA, 1994)

Saratsi, Eirini, et al., eds, *Woodland Cultures in Time and Space* (Athens, 2009)

Schlich, William, *Forestry in the United Kingdom* (London, 1904)

Sloan, Emily, *The Landscape Studies of Hayman Rooke (1723–1806): Antiquarianism, Archaeology and Natural History in the Eighteenth Century* (Woodbridge, 2018)

Stafford, Fiona, *The Long, Long Life of Trees* (London, 2017)

Steven, H. M., and A. Carlisle, *The Native Pinewoods of Scotland* (Edinburgh, 1959)

Stroud, David A., et al., *Birds, Bogs and Forestry: The Peatlands of Caithness and Sutherland* (Peterborough, 1987)

Strutt, Jacob George, *Sylva Britannica; or, Portraits of Forest Trees Distinguished for Their Antiquity, Magnitude, or Beauty* (London, 1830)

Tansley, A. G., *The British Islands and Their Vegetation* (Cambridge, 1939)

Tekić, Ivan, and Charles Watkins, 'Making Dalmatia Green Again: Reforestation at the "Horrible Edge" of Empire, 1870–1918', *Landscape History*, XLII (2021), pp. 99–118

Tittensor, Ruth, *Shades of Green: An Environmental and Cultural History of Sitka Spruce* (Oxford, 2016)

Tree, Isabella, *The Book of Wilding: A Practical Guide to Rewilding, Big and Small* (London, 2023)

Tubbs, Colin, *The New Forest: An Ecological History* (Newton Abbot, 1968)

——, *The New Forest* (London, 1986)

Vera, Frans, *Grazing Ecology and Forest History* (Wallingford, 2000)

Watkins, Charles, *Woodland Management and Conservation* (Newton Abbot, 1990)

——, *Trees, Woods and Forests* (London, 2014)

——, *Trees in Art* (London, 2018)

——, ed., *European Woods and Forests: Studies in Cultural History* (Wallingford, 1998)

——, and Ben Cowell, *Uvedale Price, 1747–1829: Decoding the Picturesque* (Woodbridge, 2012)

Whitman, Walt, *Specimen Days* [1887] (Oxford, 2023)

Williams, Mavis, *Lumber Jill* (Bradford on Avon, 1994)

ACKNOWLEDGEMENTS

I wish to thank the many friends and colleagues who have provided help, assistance and advice about trees, woods and forests over the years. Special thanks are due to those with whom I have worked on projects linked to this book. I would like to thank all the staff at Reaktion for their encouragement and very helpful advice.

Many people have provided ideas and examples used in this book and have provoked wide-ranging discussion and debate. These include Mauro Agnoletti, Dan Allen, Pantelis Arvanitis, Kostas Baginetas, Sallie Bailey, Ross Balzaretti, Luke Barley, Rebecca Batty, John Beckett, David Beckingham, Clive Brasier, Raffaella Bruzzone, Mathias Bürgi, Jill Butler, Matthew Carter, Roberta Cevasco, Hugh Clout, Harry Cocks, Amy Concannon, Fiona Cooper, Janet Cooper, Ben Cowell, Richard Cumming-Bruce, Stephen Daniels, James Davenport, Catherine Delano-Smith, László Demeter, Paul Elliott, Georgina Endfield, Robert Fish, Andrew Fox, Nicola Gabellieri, Graham Garratt, Somnath Ghosal, Ted Green, Carl Griffin, Robert Hearn, Mike Heffernan, Jake Hodder, Sophie Hollinshead, Della Hooke, Chris Ives, Melchior Jakubowski, Nuala Johnson, Matt Kempson, Kamala Khanal, Keth Kirby, Don Sandro Lagomarsini, Jack Langton, Sarah Law, Michael Leaman, Stephen Legg, Norman Lewis, Denis Linehan, Vivyan Lisewski-Hobson, Hayden Lorimer, Briony McDonagh, Simon Malloch, David Matless, Paul Merchant, Peter Merriman, Jean-Paul Métaillé, Carlo Montanari, Diego Moreno, Garrett Nelson, Mark Pearce, Valentina Pescini, George Peterken, Susan Peterken, Sandrine Petit, Carl Phillips, Pietro Piana, Pietro Piussi, Leonardo Porcelloni, Clive Potter, Enrico Priarone, Donna Radley, Stephen Radley, George Revill, Mark Riley, Graham Riminton, Jeremy Rison, Tony Robinson, Ian Rotherham, Eirini Saratsi, Susanne Seymour, Brian Short, Emily Sloan, Susan Sloman, Chris Smout, Jonathan Spencer, Adam Swain, Setsu Tachibana, Neven Tandarić, Ivan Tekić, Judith Tsouvalis, Alex Vasudevan, Lucy Veale, Frans Vera, Paul Warde, Elaine Watts, Philip Wheeler, Kate Whiston, David Whitehead, Tom Williamson and Vaclav Zaloudek.

I would also like to thank all my colleagues in the School of Geography at the University of Nottingham for their help and assistance, and the undergraduates and research students at Nottingham who provide valuable, invigorating and entertaining funds of knowledge and enthusiasm.

PHOTO ACKNOWLEDGEMENTS

The author and publishers wish to express their thanks to the sources listed below for illustrative material and/or permission to reproduce it. Some locations of artworks are also given below, in the interest of brevity.

Adobe Stock: 26 (vvoe); Alamy: 27 (michael walters 7); Alexander Turnbull Library Collections, National Library of New Zealand, Wellington, New Zealand: 3; from Henry Alford, *The Riviera: Pen and Pencil Sketches from Cannes to Genoa*, London, 1870: 44; author's collection: 4, 11 (from George Childs, *Woodland Sketches: A Series of Characteristic Portraits of Trees, Adapted for Studies for Artists and Amateurs* (London, 1839), 29, 35, 39; Bridgeman Images: 2 (The Board of Trustees of the Royal Botanic Gardens, Kew); © The Trustees of the British Museum: 12, 30; Boston Public Library: 7 (from Hayman Rooke, *Descriptions and Sketches of Some Remarkable Oaks in the Park at Welbeck, in the County of Nottingham* (London, 1790); Cleveland Museum of Art, Ohio: 9 (Mr and Mrs Lewis B. Williams collection); László Demeter: 19; The J. Paul Getty Museum, Los Angeles: 37; Library of Congress, Prints and Photographs Division, Washington, DC: 36, 41; Lloyd Library and Museum, Cincinnati, Ohio: 8 (from Jacob George Strutt, *Sylva Britannica; or, Portraits of Forest Trees, Distinguished for their Antiquity, Magnitude, or Beauty*, London, 1826); Donald Macauley: 6 (Flickr, CC BY-SA 2.0); The Metropolitan Museum of Art, New York: 47 (The Elisha Whittelsey Collection, The Elisha Whittelsey Fund, 1972); National Gallery of Art, Washington, DC: 46; National Portrait Gallery, London: 33; New York Public Library: 10; private collection: 42, 43; Science Photo Library: 16 (Duncan Shaw); The State Hermitage Museum, St Petersburg: 45; State Library of South Australia, Adelaide: 13 (part of Godson Collection); Victoria and Albert Museum, London: 31; from Robert Wallace, ed., *Eleanor Ormerod, LI. D., Economic Entomologist: Autobiography and Correspondence*, New York, 1904: 28; Charles Watkins: 5, 15, 17, 18, 20, 21, 22, 23, 24, 25, 40; Wellcome Collection: 1, 32, 48; Wikimedia Commons: 14 (Matija Volović/CC BY-SA 4.0), 34 (Imperial War Museum, London/Public Domain); Yale Center for British Art, New Haven, Connecticut: 38 (Paul Mellon collection/public domain).

INDEX

Illustration numbers are indicated by *italics*